SiC Based Miniaturized Devices

SiC Based Miniaturized Devices

Special Issue Editors

Stephen Edward Saddow
Daniel Alquier
Jing Wang
Francesco LaVia
Mariana Fraga

MDPI • Basel • Beijing • Wuhan • Barcelona • Belgrade

Special Issue Editors

Stephen Edward Saddow
Departments of Electrical and Medical Engineering,
University of South Florida
USA

Daniel Alquier
University of Tours GREMAN (UMR-CNRS 7347)
France

Jing Wang
Department of Electrical Engineering, University of South Florida
USA

Francesco LaVia
CNR-IMM sezione di Catania (Strada VIII 5—Zona Industriale)
Italy

Mariana Fraga
Universidade Federal de São Paulo (UNIFESP) Rua Talim 330
Brazil

Editorial Office
MDPI
St. Alban-Anlage 66
4052 Basel, Switzerland

This is a reprint of articles from the Special Issue published online in the open access journal *Micromachines* (ISSN 2072-666X) from 2019 to 2020 (available at: https://www.mdpi.com/journal/micromachines/special_issues/SiC_Miniaturized_Devices).

For citation purposes, cite each article independently as indicated on the article page online and as indicated below:

LastName, A.A.; LastName, B.B.; LastName, C.C. Article Title. *Journal Name* **Year**, *Article Number*, Page Range.

ISBN 978-3-03936-010-9 (Pbk)
ISBN 978-3-03936-011-6 (PDF)

Contents

About the Special Issue Editors

Stephen Edward Saddow started his involvement in SiC technology in 1992 when he started to develop optically-activated switching of 6H-SiC. He spent nearly 20 years focusing on epitaxial growth of SiC along with the development of porous SiC and novel defect reduction techniques specifically aimed at improving the material quality of 3C-SiC grown on Si substrates. For the past decade he has focused his efforts on the development of SiC for biomedical devices, first via materials studies (in-vitro and in-vivo) and more recently device prototyping. Currently he is focusing on the development of robust, implantable SiC devices for long-term operation as well as SiC-based nanostructures for the treatment of deep-tissue cancer. He is also leading a Bioelectronics Rapid Prototyping Laboratory at USF with the specific purpose of allowing for rapid translation of biomedical device research to commercial products.

Daniel Alquier is Professor and Research Vice-President at Université de Tours, doing his research in GREMAN (UMR CNRS 7347). He prepared his PhD. at the LAAS-CNRS on ultra-shallow junctions in 1998. He occupied then a position in Taiwan for PixTech-UNIPAC. Since 2000, he is working at the University of Tours and became professor in 2005. Pr. D. Alquier is the author and co-author of more than 140 papers and 6 patents. He has participated to several European and national projects. His fields of interest are wide band gap semiconductors (SiC & GaN), MEMS & NEMS and engineering for power and medical applications.

Jing Wang is a Professor and Co-Director of Center for Wireless and Microwave Information Systems (WAMI) at the University of South Florida. He got dual B.S. degrees in Electrical and Mechanical Engineering from Tsinghua University in 1999. He received two M.S. degrees in Electrical and Mechanical Engineering, and an Electrical Engineering Ph.D. from the University of Michigan in 2000, 2002, 2006, respectively. His research interests include micromachined transducers, RF/Bio-MEMS, microwave/mmWave devices, RF additive manufacturing, lab-on-a-chip/microfluidics, and functional nanomaterials. His work has been funded by grants from federal agencies (NSF, DTRA, US Army, US Air Force) and contracts from many companies totaling over $14 M. He has published more than 180 peer-reviewed papers and holds 11 US patents. He serves as the chair for IEEE MTT/AP/EDS Florida West Coast Section and he acted as the general chair or TPC chair for IEEE WAMICON Conferences in 2011, 2012, 2013, 2014 and 2020.

Francesco LaVia was born in Catania (Italy) in 1961. He graduated in Physics at the University of Catania in 1985. From 1985 to 1990 he had a scholarship at the STMicroelectronics in Catania. In 1990 he joined the CNR-IMM of Catania. In 2001 he became senior researcher and became head of the research team working on epitaxy and hetero-epitaxy of silicon carbide. He was responsible of several industrial projects and contracts and actually coordinates two European projects. In his career he has published more than 300 papers in JCR journals, 11 patents, two articles on invitation, three chapters in books and he was editor of four books. He presented several invited talks at international conferences and has co-organized several conferences and tutorials. He has been the Co-Chair of the ICSCRM2015 and Chair of the Technical Program Committee. He is member of the Steering Committee of the ICSCRM conference.

Mariana Fraga obtained her PhD in Aeronautics and Mechanical Engineering (with concentration on materials science) from the Technological Institute of Aeronautics and master's degree in electrical engineering (with concentration in Microelectronics) from the University of São Paulo (USP), Brazil. Her major research efforts are in the fields of materials science and engineering, and can be briefly summarized as follows: (i) synthesis and characterization of thin films and nanostructures, more specifically those based on silicon carbide (SiC), CVD diamond, diamond-like carbon (DLC), aluminium nitride (AlN) and titanium dioxide (TiO2), and (ii) development of micro-electro-mechanical (MEMS) sensors, microelectronic devices, solar energy conversion devices, biomedical devices, and coatings for technological applications. Currently, she is a visiting professor at the Institute of Science and Technology, ICT-UNIFESP. She also serves as Member of the Editorial Board for five international journals. She is the co-editor of the book Emerging Materials for Energy Conversion and Storage.

micromachines

MDPI

Editorial

Editorial for the Special Issue on SiC Based Miniaturized Devices

Stephen E. Saddow [1,*], Daniel Alquier [2,*], Jing Wang [1,*], Francesco LaVia [3,*] and Mariana Fraga [4,*]

1 Department of Electrical Engineering, University of South Florida, 4202 East Fowler Avenue, ENB118 Tampa, FL 33620, USA
2 GREMAN UMR-CNRS 7347, Université de Tours, 16 rue Pierre et Marie CURIE, BP 7155, 37071 Tours Cedex 2, France
3 CNR-IMM Sezione di Catania, Strada VIII 5-Zona Industriale, I-95121 Catania, Italy
4 Instituto de Ciência e Tecnologia, Universidade Federal de São Paulo (UNIFESP), Rua Talim 330, São José dos Campos 12231-280, Brazil
* Correspondence: saddow@usf.edu (S.E.S.); alquier@univ-tours.fr (D.A.); jingw@usf.edu (J.W.); francesco.lavia@imm.cnr.it (F.L.); mafraga@ieee.org (M.F.)

Received: 8 April 2020; Accepted: 11 April 2020; Published: 13 April 2020

The MEMS devices are found in many of today's electronic devices and systems, from air-bag sensors in cars to smart phones, embedded systems, etc. Increasingly, the reduction in dimensions has led to nanometer-scale devices, called NEMS (Nano-Electrical-Mechanical Systems). The plethora of applications on the commercial market speaks for itself, and especially for the highly precise manufacturing of silicon-based MEMS and NEMS. While this is a tremendous achievement, silicon (Si) as a material has some drawbacks, mainly in the area of mechanical fatigue and thermal properties. Silicon carbide (SiC) is a well known wide-bandgap semiconductor whose adoption in commercial products is experiencing exponential growth, especially in the power electronics arena. While SiC MEMS have been around for decades, in this Special Issue we sought to capture both an overview of the devices that have been demonstrated to date, as well as bring new technologies and progress in the MEMS processing area to the forefront. This Special Issue contains one review paper and nine original research papers, with both experimental and theoretical investigations, reporting the recent progress of SiC materials, processing, modeling and device technology.

The review paper of this Special Issue provides an overview of high-temperature SiC power electronics, with a focus on high-temperature converters and MEMS devices [1]. This paper mainly surveyed the research and development of SiC-based high-temperature converters as well as the existing technical challenges facing high-temperature power electronics, including gate drives, current measurements, parameters matching between each component and packaging technology.

The discussion on the original research published in this Special Issue opens with the paper on the development of a 1200V/200A full-SiC half-bridge power module by Zhang et al [2]. Their study focused on the influences of output power on the turn-on Vgs characteristics for high-power and high-frequency application. There is also a paper addressing the design and simulation of an improved 4H-SiC metal semiconductor field effect transistor (MESFET) based on the double-recessed MESFET (DR-MESFET) [3].

The use of SiC in radiation detection is the subject of two papers in this issue. Mandal et al. investigated the development of miniature 4H-SiC-based radiation detectors for harsh environment application [4], whereas Puglisi et al. reported the electrical and spectroscopic performance of an innovative position-sensitive semiconductor radiation detector in epitaxial 4H-SiC [5].

The mechanical properties of hexagonal SiC (4H- and 6H-SiC) are also discussed in this Special Issue. Ben Messaoud et al. reported the Young's modulus and the residual stress of 4H-SiC circular membranes on 4H-SiC substrates [6]. Pan et al. approached the mechanical behavior and material-removal mechanisms of single crystal 6H-SiC under the effects of abrasives by combining the morphologies of the machined surfaces and the results of nanoindentation experiments are described [7]. Chai et al. proposed a theoretical model of the critical depth of cut of nanoscratching on a 4H-SiC single crystal with a Berkovich indenter [8].

The last two articles of this Special Issue involve the synthesis, characterization and application of SiC films. Galvão et al. explored the structural and chemical properties of polycrystalline SiC films grown at room temperature on Si and aluminum nitride (AlN)/Si substrates by the high-power impulse magnetron sputtering (HiPIMS) technique [9]. Beygi et al. reported on the design and fabrication of a Michigan-style SiC neural probe on a silicon-on-insulator (SOI) wafer for the ease of the manufacture. The probe is composed of 3C-SiC, which was epitaxially grown on a SOI wafer [10]. These neural interfaces may pave the way for long-term human implants to treat such serious conditions as Parkinson's, dementia and depression, restore lost functionality due to brain damage and enable seamless integration of robotic prosthetics for patients who have suffered limb-loss.

We sincerely hope that this Special Issue on SiC-based miniaturized devices can be a valuable source of information for researchers working on this topic. We would like to thank all the authors and reviewers for their contribution and effort. We are also grateful to the editorial and production staff of Micromachines for their support. We hope that you enjoy reading this Special Issue.

Conflicts of Interest: The authors declare no conflict of interest.

References

1. Guo, X.; Xun, Q.; Li, Z.; Du, S. Silicon Carbide Converters and MEMS Devices for High-temperature Power Electronics: A Critical Review. *Micromachines* **2019**, *10*, 406. [CrossRef] [PubMed]
2. Zhang, M.; Ren, N.; Guo, Q.; Zhu, X.; Zhang, J.; Sheng, K. Modeling and Analysis of vgs Characteristics for Upper-Side and Lower-Side Switches at Turn-on Transients for a 1200V/200A Full-SiC Power Module. *Micromachines* **2020**, *11*, 5. [CrossRef] [PubMed]
3. Jia, H.; Tong, Y.; Li, T.; Zhu, S.; Liang, Y.; Wang, X.; Zeng, T.; Yang, Y. An Improved 4H-SiC MESFET with a Partially Low Doped Channel. *Micromachines* **2019**, *10*, 555. [CrossRef] [PubMed]
4. Mandal, K.C.; Kleppinger, J.W.; Chaudhuri, S.K. Advances in High-Resolution Radiation Detection Using 4H-SiC Epitaxial Layer Devices. *Micromachines* **2020**, *11*, 254. [CrossRef] [PubMed]
5. Puglisi, D.; Bertuccio, G. Silicon Carbide Microstrip Radiation Detectors. *Micromachines* **2019**, *10*, 835. [CrossRef] [PubMed]
6. Ben Messaoud, J.; Michaud, J.-F.; Certon, D.; Camarda, M.; Piluso, N.; Colin, L.; Barcella, F.; Alquier, D. Investigation of the Young's Modulus and the Residual Stress of 4H-SiC Circular Membranes on 4H-SiC Substrates. *Micromachines* **2019**, *10*, 801. [CrossRef] [PubMed]
7. Pan, J.; Yan, Q.; Li, W.; Zhang, X. A Nanomechanical Analysis of Deformation Characteristics of 6H-SiC Using an Indenter and Abrasives in Different Fixed Methods. *Micromachines* **2019**, *10*, 332. [CrossRef] [PubMed]
8. Chai, P.; Li, S.; Li, Y. Modeling and Experiment of the Critical Depth of Cut at the Ductile–Brittle Transition for a 4H-SiC Single Crystal. *Micromachines* **2019**, *10*, 382. [CrossRef] [PubMed]
9. Galvão, N.; Guerino, M.; Campos, T.; Grigorov, K.; Fraga, M.; Rodrigues, B.; Pessoa, R.; Camus, J.; Djouadi, M.; Maciel, H. The Influence of AlN Intermediate Layer on the Structural and Chemical Properties of SiC Thin Films Produced by High-Power Impulse Magnetron Sputtering. *Micromachines* **2019**, *10*, 202. [CrossRef] [PubMed]
10. Beygi, M.; Bentley, J.T.; Frewin, C.L.; Kuliasha, C.A.; Takshi, A.; Bernardin, E.K.; La Via, F.; Saddow, S.E. Fabrication of a Monolithic Implantable Neural Interface from Cubic Silicon Carbide. *Micromachines* **2019**, *10*, 430. [CrossRef] [PubMed]

micromachines

Review

Silicon Carbide Converters and MEMS Devices for High-temperature Power Electronics: A Critical Review

Xiaorui Guo [1,†], Qian Xun [2,*,†], Zuxin Li [1] and Shuxin Du [1]

[1] School of Engineering, Huzhou University, Erhuan Road 759, Huzhou, China; guoxr@zjhu.edu.cn (X.G.); XQ09086320@163.com (Z.L.); xunq520@hotmail.com (S.D.)
[2] Department of Electrical Engineering, Chalmers University of Technology, 313000 Göteborg, Sweden
* Correspondence: qian.xun@chalmers.se
† These authors contributed equally to this work.

Received: 14 April 2019; Accepted: 17 June 2019; Published: 19 June 2019

Abstract: The significant advance of power electronics in today's market is calling for high-performance power conversion systems and MEMS devices that can operate reliably in harsh environments, such as high working temperature. Silicon-carbide (SiC) power electronic devices are featured by the high junction temperature, low power losses, and excellent thermal stability, and thus are attractive to converters and MEMS devices applied in a high-temperature environment. This paper conducts an overview of high-temperature power electronics, with a focus on high-temperature converters and MEMS devices. The critical components, namely SiC power devices and modules, gate drives, and passive components, are introduced and comparatively analyzed regarding composition material, physical structure, and packaging technology. Then, the research and development directions of SiC-based high-temperature converters in the fields of motor drives, rectifier units, DC–DC converters are discussed, as well as MEMS devices. Finally, the existing technical challenges facing high-temperature power electronics are identified, including gate drives, current measurement, parameters matching between each component, and packaging technology.

Keywords: power electronics; high-temperature converters; MEMS devices; SiC power electronic devices

1. Introduction

Power conversion systems are widely employed in the industry ranging from aircraft, automotive, deep oil/ gas extraction, and space exploration where high-temperature power electronics are required [1–3]. The aircraft field is moving towards more electric aircraft with the reduction or removal of the hydraulic, mechanical and pneumatic power systems [4]. This means more electrical actuators are needed to improve the efficiency, reliability, and maintainability. Power electronic devices, such as sensors and actuators should be placed close enough to the hot engine. Among them, some need to experience the ambient temperature varying from −55 to 225 °C due to the short distance to the jet engine, and the gas turbine must operate above 350 °C [5]. In electric automotive applications, operating ambient temperature ranges from −40 °C to a very high temperature differing with various locations. For example, the coolant temperature can reach up to 120 °C at 1.4 bar, the temperatures for wheel sensor and transmission are around 150 to 200 °C, and the exhaust sensor is up to 850 °C with an ambient temperature of 300 °C [6]. In the deep oil/gas extraction, the electrical downhole gas compressor is designed to improve the throughput of gas wells. The ambient temperature is expected to reach 150 °C since the compressor should be installed close to the gas reservoir, and the system is expected to work reliably under an ambient temperature of 225 °C with the lifetime of 5 years [7]. Regarding space exploration, it is obviously a "niche" market, but it is quite challenging to develop power electronics used for this application. The surface temperature on Venus can reach up to ~460–480 °C, while on Jupiter the temperature increases with depth and pressure [8,9]. The ambient temperature reaches 400 °C at 100 bar and comes along with a very aggressive atmosphere with wind speed around 200 m/s and hydrogen-rich chemical composition. Moreover, thermal cycling can be another challenge in the space since the ambient temperature is −140 °C during the night. In these applications, it is common that electronic equipment works in a quite harsh environment, especially an extensive temperature range, and frequent deep thermal cycling [10]. Table 1 summarizes the applications requiring high-temperature power electronics in current/ future technologies.

Table 1. Applications of high-temperature power electronics.

High Temperature Application	Peak Ambient	Current Technology	Future Technology
Automotive Engine control electronics	150 °C	Bulk Si and SOI	Bulk Si and SOI
Electric/ Hybrid vehicle power management and distribution (PMAD)	150 °C	Bulk Si	WBG
Electric suspension and brakes	250 °C	Bulk Si	WBG
On-cylinder and exhaust pipe	850 °C	N/A	WBG
Turbine engine			
Sensors, telemetry, control	300 °C/600 °C	Bulk Si and SOI/N/A	WBG and SOI/WBG
Electric actuation	150 °C/600 °C	Bulk Si and SOI/N/A	WBG
Deep-well drilling telemetry			
Oil and gas	300 °C	SOI	WBG and SOI
Geothermal	600 °C	N/A	WBG
Industrial			
High-temperature processing	300 °C/600 °C	SOI/N/A	SOI/WBG
Spacecraft			
Power management	150 °C/500 °C	Bulk Si and SOI/N/A	WBG
Venus and Mercury exploration	550 °C	N/A	WBG

Note: Bulk Si, SOI, WBG, and N/A stand for bulk silicon, silicon-on-insulator, wide band-gap, and currently no available, respectively.

To be able to withstand the high-temperature environment, power electronic equipment is always designed to have active or passive cooling systems [11]. These cooling solutions include forced convection air cooled heat sink, sided cooling with liquid cold plate, micro-channel liquid cooler built into the power module, in addition to jet impingement and direct contact liquid cooling. Although large numbers of thermal management solutions have been designed to cool the power electronics and manipulate their operating temperature, these cooling solutions in oil/gas extraction applications are not efficient or effective. Besides, those applications have the typical issue associated with obviously undesired high cost, extra weight, and volume by introducing the cooling system. Moreover, due to the limited information about the actual operating environment and the actual load cycling, there is a lack of accurate thermal analysis and reliability assessment for each component [12]. Thus, failure of the cooling system can readily jeopardize and even destroy the whole electronic system. Considering these factors, it would be appealing to have electronic components capable of enduring elevated and fast varying temperatures. As a result, system reliability can be largely improved while both the upfront and operating costs can be reduced. Accordingly, the development of high-temperature power electronics is of great importance and has attracted considerable research efforts.

Currently, power electronic devices in conversion systems and micro-electromechanical systems (MEMS), such as DC/AC converters, AC/DC converters, DC/DC converters, control, ICs and sensors are mainly manufactured by silicon (Si) material [13]. The performance of Si-based power electronics has almost been driven to its logical boundary after about 60 years of tremendous development, but they still cannot offer satisfactory performance for many applications. For instance, the maximum permissible operating temperature for Si isolated gate bipolar transistor (IGBT) from Infineon is 125 °C, for Si MOSFET, the maximum obtainable junction temperature is 150 °C. Such a limited junction temperature together with switching frequency makes it unsuitable for individual application fields with required high-power, high-frequency, and high-voltage. To cater for the low-temperature tolerance, the Si IGBT is designed with complicated heat sink and maximum switching frequency at approximately 30 kHz. This structure results in cumbersome passive components, low power density, and poor dynamic performance. However, substantial improvement in power conversion systems is difficult with simple employment of fabrication technologies or Si semiconductor devices. The demerits of power electronics manufactured by Si materials are evident and thus limit industrial applications of power electronic devices [14].

Fortunately, the third-generation semiconductor materials, represented by silicon carbide (SiC) and gallium nitride (GaN), have gradually shown superior characteristics compared to Si material. Due to the wide band-gap, high breakdown field strength, high thermal conductivity, and fast electron saturation drift velocity, SiC is one of the most promising alternative materials and it is very suited for high-temperature power electronic devices [15]. Compared with conventional Si power electronic devices, commercially available SiC devices have not only better thermal stability and higher temperature tolerance but also lower switching/conduction loss. Consequently, SiC devices are quite a promising solution to converters tailored for high-temperature applications. In theory, the permissible junction temperature of SiC power electronic devices can reach as large as 600 °C, thanks to the wide band-gap of the semiconductor material, which is about three times of that of Si material [16]. The integrated power modules operating in the full temperature range are expected to be widely applied in the foreseeable future.

Research on high-temperature power electronics is not new and has been going for 20 years ago. The first review paper on this research summarized many of the same challenges we are still suffering in today's market [17]. The common conclusions have been drawn that the high-temperature power electronics cannot be developed by only one component. Although existing power converters equipped with SiC power electronic devices can operate at a top junction temperature above 150 °C [18,19], the heat-resistance attributions of high-temperature converters will disappear if gate drives cannot endure in a high-temperature environment. The associated issue for the gate drive is to be designed accordingly to match the requirement of high-temperature and high-speed capability. Furthermore, high-temperature passive components, such as capacitors, resistors, and magnetic materials, required to assemble high-temperature converters are equally as crucial as other components for reliable operation. The magnetic material can work in ambient temperatures up to 450 °C [20], but the issue comes for capacitors as most dielectric material cannot be used above 200 to 250 °C. Due to the performance of active and passive components which varies over the wide temperature range, the parameters matching should also be considered. Moreover, high-temperature packaging and integration technology are crucial for the design and development of high-temperature converters to give full play to high-temperature tolerance of SiC power electronic devices. Thereafter, there are some reviews on high-temperature power electronics, which are mainly focused on the process, packaging, ICs [21], MEMS sensors [22], and power converters [23,24], and the review on high-temperature applications is inclusive.

In view of the above aspects, a comprehensive overview of the development profiles of high-temperature power electronics based on SiC converters and MEMS devices is conducted, however, has not been presented in the available literature. This work starts with some of the most prominent applications for high-temperature power electronics, fills such a gap by discussing state-of-the-art high-temperature components technologies, including SiC material, power devices or modules, gate drives, and passive components. It is identified that advanced material techniques are essential for high-temperature power electronics. Meanwhile, several exemplary applications of high-temperature power electronics, such as motor drives, rectifier units, DC–DC converters, and MEMS devices are analyzed, and the critical factors of performance promotion for converters are highlighted. Finally, the existing challenges in further advance of high-temperature power electronics are discussed.

2. High-Temperature Components

The electrical system, from power generation, power conversion, and power transmission to all kinds of power equipment, runs in a wide temperature range. However, these systems cannot be developed without the improvement of advanced material, power electronic devices, gate drives, and passive components [20]. This section outlines the development profiles of the high-temperature components.

2.1. The Properties of SiC Material

In 1824, a Swedish scientist called J. J. Berzelius discovered the existence of SiC material, and subsequent research revealed that this material has good performance. However, SiC material had not been well-developed due to the outstanding achievement and rapid development of Si technology at that time. Until the 1990s, Si-based devices could not meet the high requirement of power electronics, such as high frequency, high voltage, high temperature, and high power density. This has once again ignited the interest of researchers in SiC material.

Since the covalent bond between carbon and silicon is stronger than that between silicon atoms, SiC material have higher breakdown electric field strength, carrier saturation drift rate, thermal conductivity, and thermal stability compared to Si material. SiC material have a variety of different crystal structures (polytypes), and more than 250 have been identified to date. Although there are many types of polytypes, only three crystalline structures exist—cubic, hexagonal, and rhombohedral. The physical properties of the current available semiconductor materials are listed as Table 2. Despite the same atomic composition in all SiC polytypes, the electrical properties differ. For instance, the band-gap for SiC ranges from 2.2 eV for 3C-SiC to 3.2 eV for 4H-SiC. Since 4H-SiC has higher electron mobility than 6H-SiC, it is a preferable option for SiC-based devices. Due to that the thermal conductivity of SiC, which is three times that of Si, and it is expected to withstand higher operating temperature for devices equipped with SiC material.

Table 2. The physical properties of the available semiconductor materials under room temperature (25 °C).

Items	SiC			Si	GaAs
	4H-SiC	6H-SiC	3C-SiC		
Band-gap (eV)	3.2	3.0	2.2	1.12	1.43
Maximum operation temperature (°C)	1580	1580	1580	600	400
Breakdown field strength (V/cm)	2.2×10^6	2.5×10^6	2.0×10^6	0.3×10^6	0.4×10^6
Maximum electron saturation velocity (cm/s)	2.0×10^7	2.0×10^7	2.5×10^7	1.0×10^7	1.0×10^7
Thermal conductivity (W/cm·K)	3~4	3~4	3~4	1.7	0.5
Electron mobility (cm^2/s·V)	980	370	1000	1350	8500
Hole mobility (cm^2/s·V)	120	80	40	480	400

However, high-purity SiC powder, which can be used to grow SiC boules, is only available from a limited number of suppliers, and is relatively expensive [25]. At present, the United States is the global leader in the production of SiC substrates and wafers, followed by Europe and Japan. The quality of SiC substrate is critical for the manufacturing of high-quality chips, and the SiC substrate constitutes a major portion of the chip cost. However, the cost of epi-growth and chips can also be reduced by the use of larger-area substrates, so manufacturers that are able to successfully fabricate 6-inch diameter SiC substrates with acceptable quality. From the Yole's report, the market size of SiC N-type wafers will increase to US$110 million by 2020 with a 21% compound annual growth rate (CAGR). With a fast growing rate of CAGR, the production of SiC-based devices will be dramatically increased.

2.2. *SiC-based Power Electronic Devices*

(1) Development of SiC devices

As early as 2001, Infineon produced the first commercial SiC Schottky barrier diode (SiC SBD) with characteristics of high blocking voltage, better thermal stability, and hardly any reverse recovery time. This paved the way for the development of SiC power devices in the field of power electronics. Since then, more discrete devices and power modules have gradually come out [26]. Figure 1 shows the milestones of the development process of commercialized SiC semiconductor devices. Until 2014, GeneSiC and Micross components have sold SiC bipolar junction transistor (BJT) with junction temperature up to 210 °C. At the research and development level, the operating junction temperature of SiC-SBD can reach up to 300 °C, and the performance of SiC positive-negative (P-N) diode under the temperature of 600 °C has also been verified.

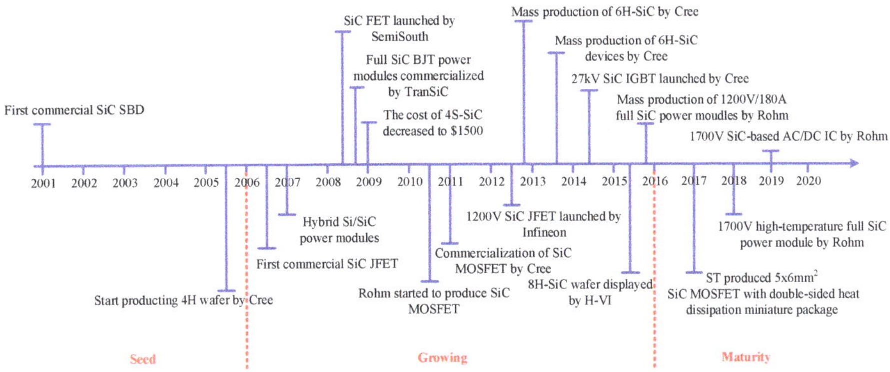

Figure 1. The milestones of the development process of SiC power electronic devices.

As the most market-oriented SiC devices at this stage, SiC MOSFETs have a fast switching speed and low on-resistance. In 1987, Palmour et al. from NCSU in USA developed the world's first high-temperature depletion layer N-channel MOSFET. Subsequently, Brown et al. from GE integrated a simulated operational amplifier (OPA) using depletion MOSFET, and it can work at 300 °C. In 2011, Purdue University reported a SiC CMOS digital integrated circuit with the maximum temperature of 350 °C, but as the temperature continues to increase, the gate leakage current will increase rapidly. Studies have shown that the long-term reliability of the gate oxide structure of SiC MOS is not good, especially at high temperature, this issue is exacerbated. At present, commercial SiC MOSFETs can operate up to 200 °C [27].

Unlike SiC MOSFETs, SiC BJTs, with high reliability, are very suitable for high-temperature conditions. However, the disadvantage is that a continuous and stable driving current is required to cause a large loss, and the current gain decreases as the temperature increases, and then the driving loss is further increased. The commercialized 1,200 V SiC BJT produced by GeneSiC can withstand temperature up to 210 °C, which is the highest level in the market. Actually, SiC junction field-effect transistor (JFETs) has developed since the 1990s, and the first commercial SiC JFETs came out around 2006. In general, a lateral channel structure or a vertical trench structure is employed in a SiC JFETs. It shows that Infineon uses lateral channels, while Semi South mainly uses vertical channels. SiC JFETs produced by Semi South and packaged by Micross are resistant to temperature up to 200 °C. SiC JFETs is currently being studied by NASA Glenn Research Center, Rutgers University and Caesar Western Reserve University, and it is reported that SiC JFETs can operate reliably for 521 h at 460 °C. In 2016, NASS reported that SiC JFETs can operate for 25 h at 727 °C. Figure 2 compares the highest tolerated temperature for commercial power electronic devices at the current stage [28]. The advanced semiconductor materials are becoming the new choice for high-temperature power electronics.

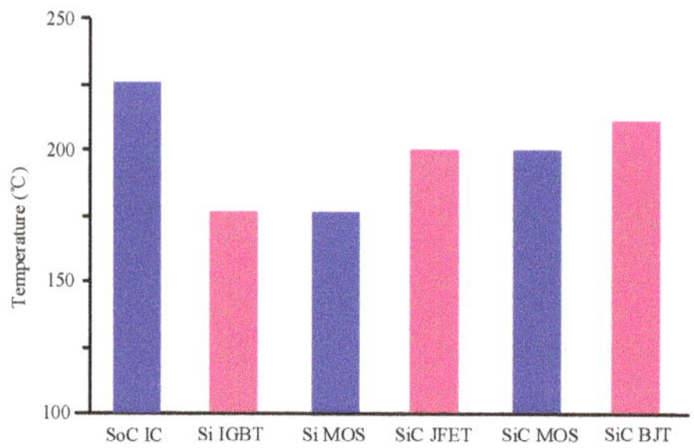

Figure 2. The maximum operating temperature of SiC power electronic devices.

Theoretically, SiC devices, with wide band-gap, can allow a very high voltage and high operating temperature. However, the thermal capability of all materials has not reached the same technological maturity. The maximum operating junction temperature for most commercial SiC devices is only up to 210 °C. Tennessee University has developed the 1.2 kV/100 A SiC JFET power module operating at 200 °C. In [29], a 1.2 kV/60 A SiC MOSFET phase-leg power module with the optimized internal layout is presented for an operating frequency of 100 kHz and junction temperature of 200 °C. In [30], a SiC power module with a junction temperature of 250 °C is presented for military hybrid electric vehicle applications, which is designed as half or full bridge structure. Some discrete devices and ICs are demonstrated laboratory level to operate above 500 °C for a short while. Reference [31] shows the characteristics of MOSFET fabricated on β-SiC thin films, which can operate at the temperature of 650 °C. The research on material and fabrication of SiC devices is still ongoing to develop the high-temperature commercial SiC devices and modules.

(2) Fabrication of SiC devices

For an example of processing of SiC MOSFETs, wafer sizes and material quality for SiC have improved over time. The main difference between the processing of Si and SiC wafers is the temperature range, shown in Reference [32] for details. Since the strong bonds between silicon and carbon need more energy for the growth of material, post-annealing of damaged material after ion implantation, bond breaking during thermal oxidation or contact alloying. A simplified SiC MOSFET process flow in Figure 3 starts with ion implantation, field oxide formation, and polycrystalline silicon gate stack alignment. For SiC MOSFETs, the channel mobility improves slightly at higher temperature, and there is a maximum voltage rating that strictly limits the current drive and on-resistance. For ICs operating at high temperature, the condition will approach that of accelerated lifetime testing. Still, threshold voltages have to be designed with some safety margin at the highest temperature expected, where it is reduced by about 1 V with the rise of 100 °C.

At present, packaging technology is a big issue for design and fabrication of SiC power electronic devices or modules for high-temperature applications. Most packaging solutions are developed for mild ambient, and permissible junction temperature of SiC devices is far below the theoretical value. Packaging material and technology are critical factors for the further increase of the operating temperature. The future high-temperature converters require higher power density, which requires that the heat dissipation conditions would be as simple as possible, and the packaging technology should also adapt to the high-temperature situation.

Ohmic contact and Schottky interface are critical factors that may limit the devices in the application of high temperature. High operating temperature could result in a diffusion process in the contact layer and a reaction between the contact components, which may result in changes in contact properties during high temperature operation, as well as degradation of the devices. In general, low contact resistivity should be maintained to decrease the voltage drop, but to get the low resistivity of ohmic contact is difficult for SiC devices due to the difficulty in doping and, in the case of p-type materials, due to the high electron affinity and high width of the band-gap. In addition, with the increase of temperature, Schottky barriers between metal and semiconductor become larger, which makes ohmic contact on SiC more difficult. By contrast, with Si and GaAs devices, the operating temperature is limited by the electronic properties of the semiconductor material; but the maximum operating temperature of SiC devices is limited by the stability of the contacts. Some device parameters such as response time and output power, depend strongly on the ohmic contact resistivity and its stability at high operating temperature. Therefore, the thermal stability of ohmic contacts and Schottky interfaces at high operating temperature is considered the critical factor for determining their power application.

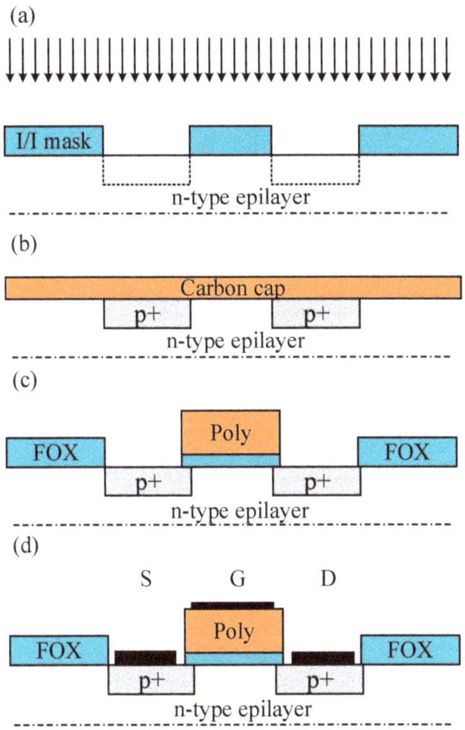

Figure 3. Simplified non-self-aligned SiC metal oxide MOSFET (**a**) Ion implantation of p-type dopants (arrows) into source and drain (dashed boxes) is performed with an ion implant hard mask; (**b**) during ion implantation annealing, the surface has to be protected with a carbon cap, which prevents Si out-diffusion and surface roughening even at 1,800 °C; (**c**) the field oxide (FOX) and polycrystalline silicon gate (poly) stack is aligned with global alignment marks (not shown), and some overlap must be allowed; (**d**) the metal must be patterned before annealing at ~800–900 °C to form the metal silicide (otherwise the reaction will occur with SiO_2).

Device applications are presented as discrete devices and power modules, and wire-bonded package, shown in Figure 4, is commonly used in high-temperature converters. The substrate, base plate, die attach and heat sink are included, and the following will analyze the physical structure and composition material of SiC power module.

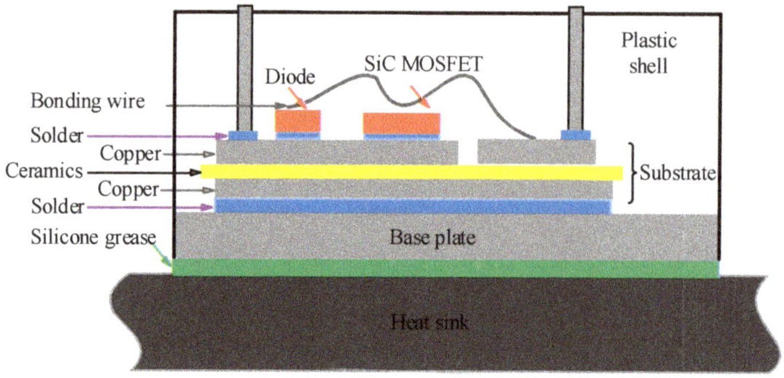

Figure 4. The package of SiC power module.

- Substrate

The substrate, as the backing of SiC chips, must meet the material requirement with characteristics of excellent thermal conductivity, high mechanical strength, high flexural strength, and the similar coefficient of thermal expansion (CTE) with SiC. The direct-bond-copper (DBC) substrate presents a sandwich structure, which consists of two layers of copper, lying on the top and bottom, and with insulation ceramics in between. The available materials for insulating ceramics could be quantized as Al_2O_3, AlN, BeO, and Si_3N_4, and Table 3 shows the thermal mechanical characteristics. As it can be seen, the CTE of AlN is close to that of SiC material with the value of approximately 4–6 ppm/°C, and the mechanical stress resulted by heat expansion can be significantly reduced by the uniformity of parameters. Moreover, the thermal conductivity of AlN is highly relative to two other materials like Al_2O_3 and Si_3N_4, which will significantly promote the cooling capacity of power modules. Therefore, AlN is one of the most appropriate choices for the substrate to encapsulate SiC power modules, and the relevant power modules can endure in the high-temperature environment which is above 250 °C [33].

Table 3. The thermal mechanical characteristics of materials for the substrate.

Ceramic Materials	96% Al (Al_2O_3)	99% Al (Al_2O_3)	AlN	BeO	Si_3N_4
Dielectric field intensity (kV/mm)	12	12	15	12	10
CTE (ppm/°C)	60	7.2	4.5	7.0	2.7
Thermal conductivity (W/m·K)	24	33	170	270	60
Structural strength (g/cm^3)	3.5	4.0	2.6	4.0	5.0
Flexural strength (MPa)	317	345	360	250	850
Tensile strength (MPa)	127	207	310	230	17

- Base plate

The base plate, as the foundation and heat conductor of power modules, also has the strict material requirement, which should have characteristics of excellent thermal conductivity and similar CTE with the substrate [34]. Table 4 shows the thermal mechanical attributes of possible materials for the base plate. CTE of Cu and Al are much higher than that of AlN (only 4.5 ppm/°C), while CTE of metal matrix composites with Mo and Cu is close to that of AlN. The thermal conductivity of metal matrix composites makes these materials have excellent heat dissipation property. With these advantages, the metal matrix composite is an excellent choice for power modules to endure in a high-temperature environment. Due to the adjustable CTE from 6.5 to 9.5 ppm/°C, and relatively high thermal conductivity with the range of 170–200 W/m·K, AlSiC is also an excellent material to constitute base plate.

Table 4. The thermal mechanical attributes of materials for the base plate.

Materials	100 Cu	100 Al	85 Mo/15 Cu	75 Mo/25 Cu	65 Mo/35 Cu
CTE (ppm/°C)	17.8	26.4	6.8	7.8	9.0
Thermal conductivity (W/m·K)	398	230	165	185	205
Density (g/cm3)	8.90	2.70	10.01	9.87	9.74
Young modulus (MPa)	128,000	70,000	274,000	274,000	274,000
Yield strength (MPa)	210	50	/	/	/

- Die attach

Die attach, also known as die bonding, is the process of connecting or bonding a die or chip to a substrate, package or another die. This process can happen in many forms and can be applied in many ways, and differences between them are entirely dependent on the user's desire. Various options are discussed in [20], including high lead solder alloys, lead-free solders, high-temperature reflow soldering, nano-silver paste sintering, and transient liquid-phase bonding. Transient liquid-phase bonding is capable of bonding at low temperature and servicing at high temperature, which is a significant trend in the field of high-temperature packaging for SiC power modules. Due to the disadvantages of time-consuming nature and required special equipment, some research on high-throughput solutions is under way [35].

Solder is required to have characteristics of a high melting point, strong bonding strength, excellent thermal conductivity, and electrical conductivity. It can efficiently conduct the heat generated by switching losses to a radiator in time. Thus, it dramatically improves the capabilities of SiC power modules to operate reliably under the condition of high frequency and high-temperature [36]. Some readily available lead-free alloys fall within the category of the high melting point, with melting points well above 220 °C, which are listed in Table 5. The lead-free alloys are a good option for high-temperature die attach. As can be seen, Au20Sn80 with the highest melting point of 280 °C is suitable for high-temperature soldering and packaging. Gold-based solders such as AuSn, AuGe, and AuSi, have a good reputation because of their melting points of 280 °C, 365 °C, 363 °C or so, respectively, but the gold-based solders are costly and not suitable for mass production. Other alloys such as zinc and bismuth alloys have been studied for their potential as well [37,38]. Silver sintering has better long-term reliability compared to leaded-free solders, but the challenge of microstructure evolving or coarsening at elevated temperature is a hindrance.

Table 5. The thermal mechanical characteristics of materials for die attach.

Solder	Sn96.5 Ag3.5	Sn95 Sb5	Pb75 In25	Sn10 Pb88 Ag2	Sn10 Pb90	Au20 Sn80
Melting point (kV/mm)	221	235	240	267	275	280
Boiling point (ppm/°C)	221	240	260	290	302	280
Density (W/m·K)	7.5	7.25	9.97	10.75	10.75	14.51
Thermal conductivity (W/m·K)	33	28	18	27	25	57
CTE (ppm/°C)	30	31	26	29	29	16
Electrical conductivity (S/m)	16	11.9	4.6	8.5	8.9	/

- Heat sink

Generally, the heat sink is joined to the power module by silicone grease, and in this way, the heat generated by all kinds of losses dissipates to the air. The thermal resistance of silicone grease occupies 50% in the whole module, and thus thermal interface materials (TIM) become one of the critical factors to significantly reduce the radiator volume and increase converter power density [39]. Moreover, the grease will dry out and age as time goes on. Therefore, some innovative solutions to reduce the thermal resistance and optimize the power dissipation should be developed.

TIM research has focused on three primary areas of using high-performance fillers, studying the micro and nanoscale TIM, and developing carbon-allotropes-based TIM to enhance performance [40–42]. Carbon nanotube fillers are the most promising of all the carbon allotropes, and have become front research topics in TIM which can decrease thermal stress caused by CTE mismatch and can keep chemicals table over a wide range of temperatures.

The high-performance cooling methods are jet impingement cooling, spray cooling, and micro channels, and all of them can be designed as two-phase flow with the increased heat transport capability [43,44]. A model of single and two-phase liquid cooling using micro channels was established for GaN-on-SiC field effect transistor (FET) [45]. Furthermore, an additional method is the direct-immersion cooling technique, where the coolant liquid is in direct contact with the power electronics device. However, the notable problems of coolant leakage and compatibility requirements result in preference for indirect cooling.

2.3. High-temperature Gate Drives

In ambient temperatures above 150 °C, a real challenge is to seek a proper gate drive circuit to match with attributions of heat resistance and high frequency of SiC power electronic devices. In theory, the junction temperature of SiC power electronic devices can reach up to 600 °C, which are commercially available for high-temperature applications. However, the integrated drive circuits for SiC devices are still in an early stage of research and development.

Based on the silicon-on-insulator (SOI) technology, a buried insulator layer in SOI structure shown in Figure 5 can effectively reduce the leakage current in high-temperature operation, improve the latch-up immunity, and suppress the threshold voltage variation to the temperature. The SOI-based integrated circuits can successfully operate at the temperature range from 200 °C to 300 °C [46], which is much higher than that of conventional bulk Si devices.

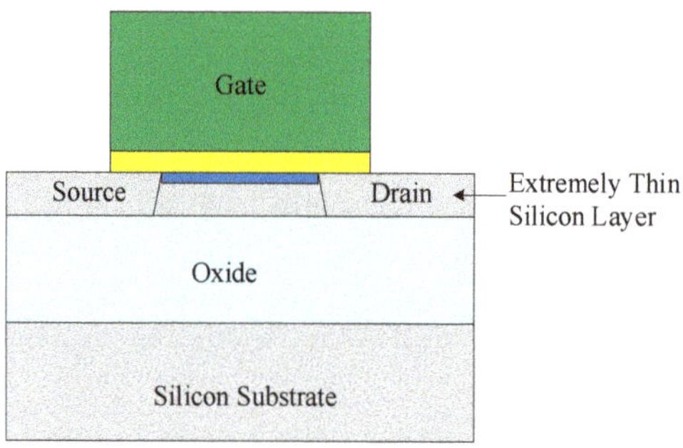

Figure 5. A buried insulator layer in a fully depleted SOI structure.

Compared to the SOI-based integrated circuit technology, the SiC-based integrated circuit technology is more attractive to extremely harsh environment [47]. However, the development of semiconductor technology presents many difficulties. Researchers have spent nearly half a century producing the first commercial SiC devices since the excellent characteristics of SiC material were discovered. However, the SiC-based integrated circuit remains at the stage of laboratory investigation, and is therefore unable to access the market in a short time. Regarding high-temperature gate drives of SiC devices, SiC-based integrated circuit technology is not mature enough yet, and Si-based integrated circuit technology is unavailable for high-temperature applications. Instead, Si-based discrete device technology may be a feasible solution when compensating temperature drift. In addition, the SOI-based integrated circuit is available for high-temperature applications when ignoring the high manufacturing cost [48].

High-temperature gate drives are required to have the input signal isolation, wide duty cycle range, static turn on/off, and low propagation delay time. Currently, no commercial opt couplers can be used in a high-temperature condition. As a result, the transformer becomes the best choice for high-temperature isolation of the gate drive. For a developed SiC MOSFET phase-leg power module, an input signal isolator is used to build an SOI gate drive chip. Even though SOI die, printed circuit board (PCB) and passive components can accommodate the high-temperature environment above 150 °C, but the SOI gate drive board cannot withstand such temperature. Its bottleneck is the commercial isolator, whose maximum operating temperature is 105 °C. Air-core transformer isolators for the SOI-based integrated circuit are under active study, with the hope to reduce the volume of natural chip-level integration and to eliminate the high-temperature ageing effects of magnetic materials [49].

As the SOI technology is well developed, companies, such as CISSOID, XREL, and Honeywell, have developed the high-temperature gate drive circuits. CISSOID provides an isolated gate drive circuit integrated with high-temperature SiC MOSFET devices in the shape of the intelligent power module (IPM), which dramatically reduces the effect of the parasitic parameters and allows the maximum ambient temperature up to 225 °C [50]. The gate drives manufactured by XREL can operate at the ambient temperature of 230 °C. The SOI technology can also be used to develop high-temperature drive ICs, but the manufacturing process is complicated and associated with the high cost. The gate drive ICs based on SiC technology can operate reliably at the temperature of 400 °C, and it has been reported in [51–54]. However, this technology is not mature and commercialized yet.

Due to the high cost of SOI ICs and SiC fabrication technologies, an alternative way to develop high-temperature gate drives is the utilization of commercial-off-the-shelf (COTS) discrete transistors and diodes. With the COTS discrete component, the operating temperature of gate drives can reach up to 180–200 °C [55]. The main drawback of this kind of gate drives is the large propagation delay due to the high number of SOI ICs, and the protection features of desaturation and under voltage lockout (UVLO) are not included. To solve these issues, a COTS gate drive is proposed with the integrated circuit of overcurrent and UVLO, which can operate under the ambient temperature of 180 °C [56]. The proposed COTS gate drive shows better performance than the commercial gate drive (EVK-HADES 1210) produced by CISSOID in the aspects of propagation delay and total power consumption. Figure 6 shows the prototype of the COTS gate drive and the test bench. In Figure 6a, the numbers shown from 1 to 5 represent on-state voltage monitoring diode, driving buffer stage, overcurrent detection circuit, UVLO with control logic, and isolated transformer, respectively.

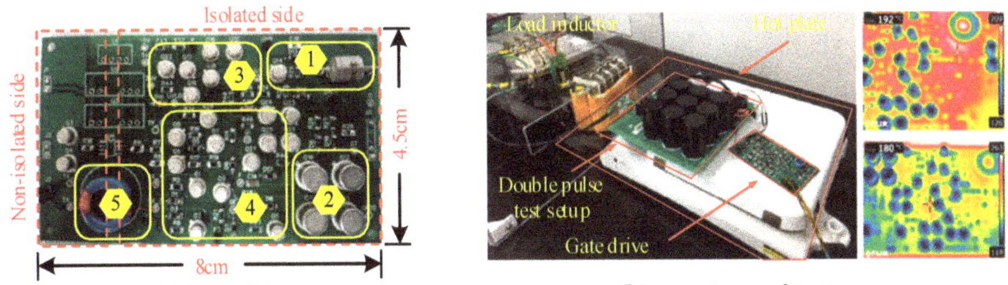

(a) Gate drive (b) experimental test setup

Figure 6. Developed high-temperature gate drive and the experimental set setup.

2.4. High-temperature Passive Components

(1) Capacitors

Capacitors are an irreplaceable component in the power electronics, and they need to be selected carefully for converters, since capacitance, insulation resistance, reliability, and service lifetime will be reduced to some extent with the increase of temperature [51]. The temperature dependency of these characteristics is related to capacitor structures and materials. Among various dielectric materials, only a few are available for high-temperature operation, such as ceramic, mica, and tantalum [57]. Mica-10 shows prominent characteristics even at 200 °C with an energy density of 11.27 J/cm^3 and efficiency of

94.7% at 500 MV/m, which is 30 times higher than the well-known polymer PI for high-temperature applications [58]. The tantalum capacitor is one option for high-temperature applications, and it can reliably work at elevated temperatures above 200 °C [59]. Considering the low capacitance of the mica capacitor and low voltage ratings of the tantalum capacitor, the ceramic capacitor is a preferable option for high-voltage and high-power power electronics.

C0G and X7R ceramic capacitors are rated for high-temperature operation. The capacitance of C0G decreases by 1% when the temperature varies from 25 °C to 250 °C. Although C0G capacitors present excellent temperature stability, capacitances are too low to meet the requirement of the high-power system. For X7R capacitors, capacitance is in the range of microfarad, which is almost an order of magnitude larger than C0G capacitors. However, the X7R capacitors present wicked capacitance stability, and the capacitance will be reduced 42% from 25 °C to 250 °C coming along with an excessive capacitance variation with operating voltage [60]. Moreover, the insulation resistance is only 210 kΩ at 200 °C, which is far below the 3 GΩ of C0G capacitors. This makes X7R capacitors unsuitable for most signal processing applications.

Alternatively, the stacked ceramic capacitors with XHT dielectric from Presidio components, Inc. present improved capacitance stability when DC bus voltage is very high, as shown in Figure 7. A capacitance reduction of 50% from 25 °C to 175 °C is quantified. This kind of capacitors is more to cracks when used under the conditions of shock and vibration. Thus, the de-rating and reliability should be considered when designing capacitors for a high-temperature converter. With the new materials and fabrication technologies, Si capacitors exhibit the promising temperature resistance up to 250 °C [60]. Table 6 shows some commercially manufactured capacitors for high-temperature applications, and few comments are also made in this table.

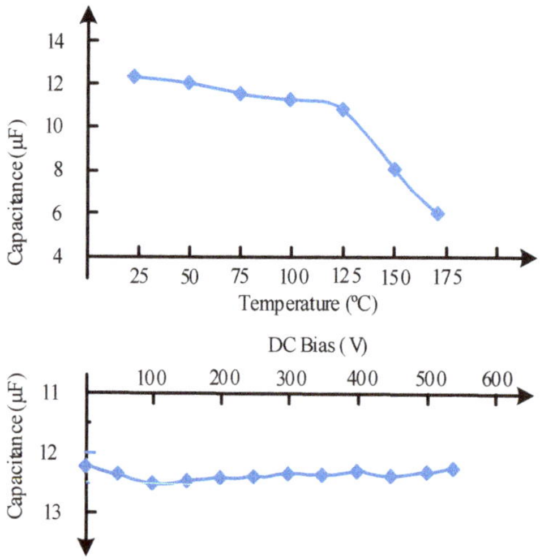

Figure 7. Capacitance stability with temperature and DC link voltage for the XHT stacked ceramic capacitors from Presidio.

Table 6. Commercially manufactured capacitors for high-temperature applications.

Capacitor type	Packaging	Max. Temperature	Comments
Solid tantalum	SMT and through-hole	230 °C	High capacitance value; voltage degradation 60–80% at 230 °C
Electrolytic wet tantalum	Axial through-hole	200 °C	High capacitance value; most through-hole
Stacked ceramic	SMT-chip stacked gull-wing	260 °C	Poor stability above 175 °C for X7R, but higher capacitance density than C0G

(2) Resistors

Resistors represent the basic component providing protection, ballasting, sensing, feedback, and signal attenuation in power electronics. Signal processing is the main contribution of resistors, in addition to high reliability and stability, which are required in such systems. Resistor manufacturers have made great effort to develop high-temperature resistors. Table 7 summarizes some commercially manufactured resistors with some short comments [16].

Table 7. Commercially manufactured resistors for high-temperature applications.

Resistor Type	Packaging	Max. Temperature	Comments
Metal foil	Through-hole, surface-mount device (SMD)-chip, flip-chip	240 °C	High precision
Metal oxide	Through-hole	275 °C	General purpose, power resistors, good frequency characteristics
Thin film	SMD-chip, flip-chip	275 °C	Compact, low TC, high stability
Thick film	SMD-chip, flip-chip	300 °C	General purpose, wide resistance range
Wire-wound	Through-hole	350 °C	High surge capability, precision power

Metal foil resistors have high precision and stability in harsh environments, but the maximum operating temperature is limited to 240 °C due to the bondable chip, wire-wound resistors show quite good high-temperature characteristics, but they are not suitable for high frequency, and they act as inductors at high frequency. Thin film resistors seem to be an economical way with small size and good performance, but they are not suitable in overload conditions, whereas thick film resistors show superior overload characteristics [61].

(3) Magnetic elements

For inductors and transformers, it is necessary to take into consideration the high-temperature characteristics and insulation properties of windings and magnetic cores. High-temperature-resistant enameled wires and Kapton electrical tape are readily available above 200 °C. As for the magnetic cores, high-temperature characteristics are affected by Curie temperature, coating materials, and power losses. The optimal operating point for most magnetic material is approximately 100 °C for most magnetic materials, and the efficiency will be reduced with the further rise of temperature.

The magnetic elements with rated operation temperature above 200 °C mainly involve three types of magnetic cores such as ferrite cores, power cores, tape wound, and cut cores, which are usually utilized for power supply, signal isolation and current detection in high-temperature converters. However, the operating temperature will affect their magnetic properties, and the B/H curve will drift with the temperature variation, which will lead to a significant detection error, especially for measurement sensors. Therefore, a compensation algorithm is needed to improve the detection precision. Reference [62] shows a strong correlation between the temperature variation and the magnetic characteristics of Mn–Zn Ferrite materials, the magnetic characteristics should be assessed when these components work in such thermal conditions. Ferrite cores with Curie temperature of 350 °C have been commercially available so far, and they can work reliably up to 250 °C.

3. High-Temperature Converter and MEMS Devices

With the commercialized development and mature application of SiC power electronic devices, high-temperature converters and MEMS devices have brought and will bring broad prospects in electric vehicles, more electric aircrafts, deep-earth oil explorations, and geothermal energy exploitations.

3.1. SiC-based Motor Drive

High-temperature motor drives are required for the flight control actuators and fuel pumps of the aviation system as well as for the control units of electric vehicles, which will face the challenge to meet the requirement of higher efficiency, higher power density, and higher reliability in harsh operation condition. Take an example of a new generation of high-power electric vehicles; the motor drive system will undergo a change from traditional industrial grade to automobile industrial grade. United States Department of Energy draws up the development goal for hybrid electric vehicle (HEV) by 2020, the dramatic advancement will be developed. The power density of power electronic equipment will be higher than 14.1 kW/kg, and volume will be less than 13.4 kW/L, efficiency will be higher than 98%, the price will be lower than 3.3 $/kW [63]. SiC power electronic devices make motor drives have the ability of high-temperature operation, which plays an essential role in goal achievement.

Figure 8 shows the electrical connection diagram of a three-phase inverter and motor, where a selection of power electronic devices is full SiC-MOSFET, full SiC-JFET, full SiC-BJT, or SiC/Si (e.g., SiC-SBD/Si-IGBT) hybrid devices. As the anti-parallel diode, SiC-SBD is hardly any reverse recovery time and is not affected by temperature variation. The radiator volume of a 2.5 kW motor drive based on SiC-SBD hybrid devices is reduced by 2/3 when compared with the radiator of the motor drive based on Si diode [64]. In [65], a three-phase air-cooling inverter is designed by using full SiC-JFET devices, the motor drive can work in the high-temperature environment of 200 °C, and the output efficiency with 18 kW power rating reaches up to 98.2%.

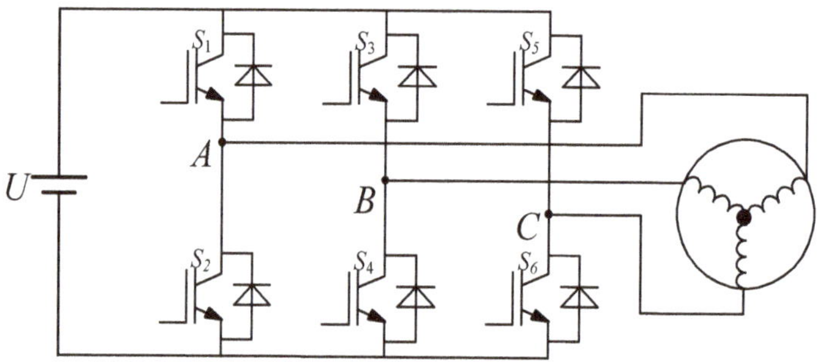

Figure 8. Electrical connection diagram of three-phase inverter and motor.

APEI Company has developed a 4 kW high-temperature motor drive by using the SiC-JFET multichip power module (MCPM) and SOI-based integrated circuit technology. The core of control electronics is a high-temperature microcontroller (SOI-based MOS HT83C51) produced by Honeywell Inc., outputting low power signals. Thus, an amplifying circuit provides a connection between the microcontroller and the power module, and the amplifying circuit is constituted of the high-temperature operational amplifiers and high-temperature passive components. A high-temperature transformer is designed by APEI to isolate the digital controlled circuit from the power circuit. Figure 9 shows the mentioned MCPM with high-density power electronics designed by APEI [66]. The MCPM approach is one of the packing strategies for power electronics, and the main idea is that the control and power circuitry components are integrated together into a single compact power module. AlN DBC material for power substrate and AlSiC material for heat spreader connected to the DBC substrate are selectable due to their excellent thermal conduction capabilities and a close CTE match. Multilayer polyimide for PCB is available for high-temperature applications since the glass transition temperature (Tg) is as high as 260 °C. An implementation of MCPM in a 4 kW 3-phase motor drive shows the SiC-JFET can operate at a junction temperature of 250 °C, Figure 10 shows the prototype of the 3-phase motor drive and the 3-phase multichip power module. Based on these state-of-the-art technologies, the motor drive module can operate in a high-temperature environment. Meanwhile, the converter has a substantial increase in respect of efficiency and power density.

Figure 9. Multichip power module with high-density power electronics designed by Arkansas Power Electronics International.

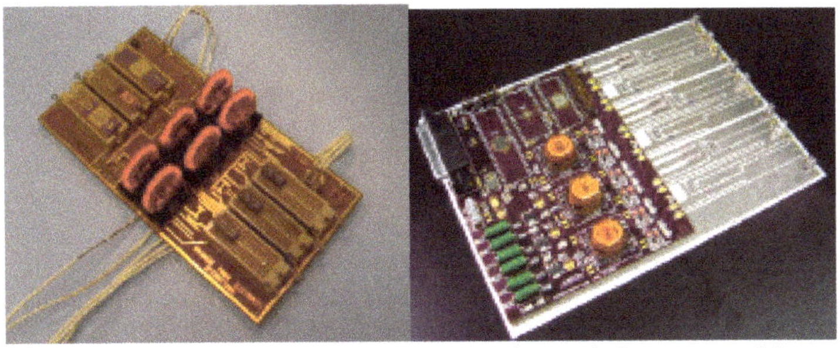

(**a**) 3-phase motor drive prototype (**b**) 3-phase multichip power module

Figure 10. A prototype of 3-phase motor drive by using multichip power module (MCPM).

3.2. SiC-based Rectifier Unit

The DC secondary power source is necessary for more electric aircraft in which the main power supply is an AC power source [67]. Here, the transformer rectifier or three-phase AC–DC converter can be used for rectifying 115 V AC bus voltage into 270 V DC, which supplies power to nitrogen generator system, engine starting system, environmental control compressor, hydraulic pump and other loads [68]. In a diode rectifier circuit, SiC-SBD with little reverse recovery current is used to reduce the switching losses, heat radiation requirements and high-temperature limitations for aviation rectifier unit. NATO is planning that SiC power electronic devices are applied to meet needs for converters to operate in a high-temperature environment in all-electric combat vehicles. Then, a 30 kW rectifier bridge is designed for the automobile application with 200 °C operation temperature.

For large-power high-temperature converters, the efficiency requirement is especially crucial except for reliability requirement in a harsh environment. The high-temperature converter with SiC power electronic devices can have a significant increase in efficiency due to the substantial reduction of switching losses. Figure 11 shows the main types of losses in a voltage doubling rectifying circuit and the efficiency comparison of the rectifier with Si IGBT and SiC MOSFET in same voltage level and control mode. The efficiency of SiC devices rectifier is 2.6% higher than that of rectifier equipped with Si devices. In addition, superior electric characteristics with low on-resistance at high-temperature and simultaneous high-speed switching conditions are also come with SiC devices rectifier.

Figure 12 shows the hardware configuration diagram of the high-temperature three-phase AC–DC converter with the integration of 1 SiC-JFET and 7 SiC-SBDs, which can be applied to more electric aircrafts with the 270 V DC output, and it can also operate reliably in 200 °C ambient environments. Each component in the rectifier system should be carefully designed. Regarding SiC power modules, the mismatch of parasitic parameters, the variance of device property, and nonidentity of junction temperature can result in the unbalanced electro-thermal stress. The imbalance of multichip in parallel can be reduced by optimizing the layout of a direct bonding base, adding the low series resistance and coupling inductance, designing the feedback control of the drive circuit [69]. Here, the SiC power module using the flat-packaged structure has advantages of small parasitic parameters, flexible line layout and, double side cooling characteristics. In switching individual tests, the parasitic inductance is reduced by 14 nH compared to the pin-packaged structure, which makes the drain-source spike voltage decrease from 295.7 V to 279 V and further lowers switching losses [70]. In temperature detection, the temperature rise of grid resistance and the voltage regulator is apparent, but the temperature rise is not more than 10 °C from the environment temperature, which benefits from the lower power losses and excellent heat-sinking capability of the high-temperature AC–DC converter.

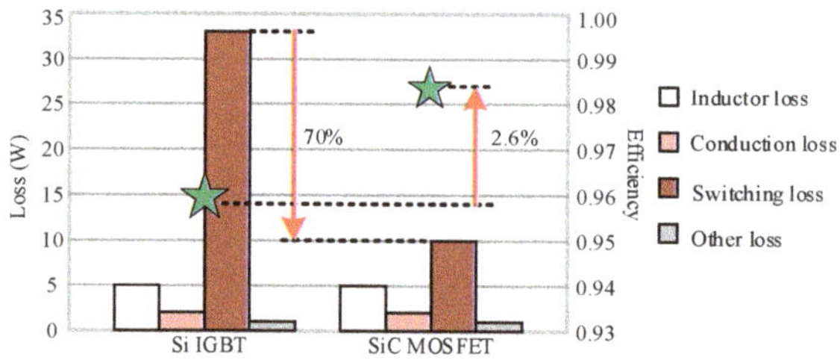

Figure 11. Loss and efficiency comparison of the rectifier with Si IGBT and SiC MOSFET.

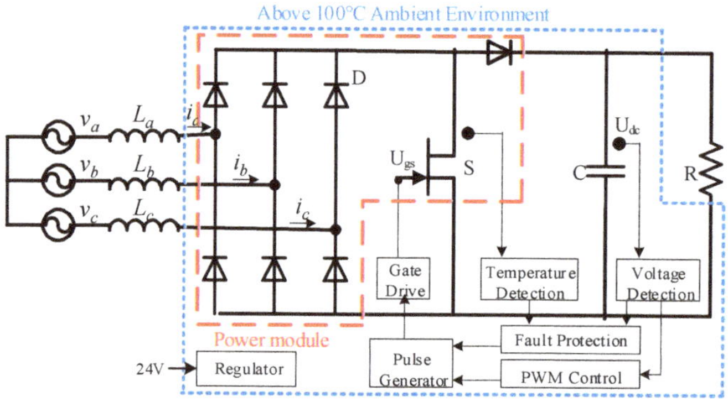

Figure 12. Hardware configuration diagram of the high-temperature three-phase AC–DC converter.

In [71], Virginia Tech developed a 15 kW 650 V dc/230 V ac three-phase rectifier with interleaving structure by substitution of all Si devices with SiC JFETs, and the SiC power modules in the rectifier can operate at the junction temperature of 250 °C. With the volumetric power density of 6.3 kW/L, it successfully achieved a 2 kW/L target in more electric aircraft. It gives a detailed design for each component, including the active component, passive component, and the system. Figure 13 shows the drawing and the prototype of the rectifier system.

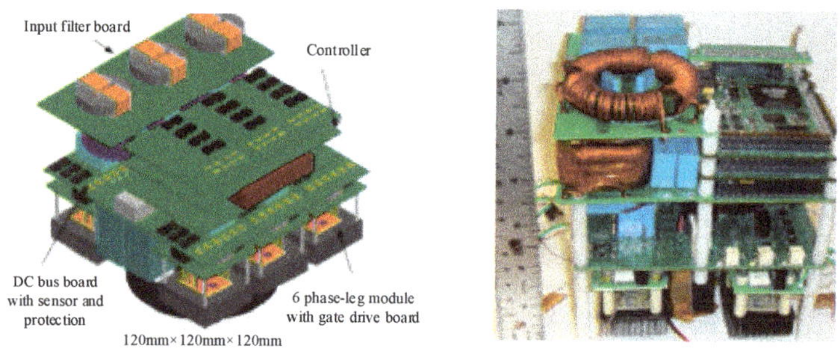

Figure 13. Conceptual drawing and the assembled prototype rectifier system.

3.3. SiC-based DC–DC Converter

DC–DC converters are widely used in the electric vehicle, more electric aircraft, and renewable energy. For example, high power HEV, the electric drive system is made up of a storage battery, DC–DC converter, inverter, electric machine, and control circuit, in which DC–DC converter plays a role in boosting the DC voltage for the post inverter unit. The permissible operating temperature for main components in HEV is listed as 120 °C for the motor, 200 °C for a turbocharger, 200 °C for throttle, 145 °C for the gearbox, 175 °C for driving chain, and 650 °C for exhaust pipe [72]. Due to the electrification and electromechanical integration of HEV, as well as the limitation of self-cooling capacity, these harsh conditions give power converters characteristics of reliability when operating in a high-temperature environment. Another example is when the DC–DC converter is applied to the generator control unit (GCU) in more electric aircrafts. The converter provides different levels of DC voltage for the various functional modules of GCU. The high voltage levels reach up to tens of thousands of volts, which are mainly used for communication, radars, transmitters of electronic warfare equipment, and a variety of cathode ray tube displays. The low voltage levels are classified as 24 VDC, ±12 VDC, ±5 VDC, ±6.3 VDC, and ±3.3 VDC. Due to the development tendency of the control system from traditional centralized engine mode to the distributed mode, the GCU would be placed closer to the generator, which makes converters operate in a harsh environment (– 55 to + 200 °C).

Figure 14 shows the circuit diagram of the boost DC–DC converter with SiC-MOSFET and SiC-SBD. Usually, Si-MOSFET has a large junction capacitance, which defines the switching frequency up to 100 kHz. Si-IGBT has current lag which limits the switching frequency up to 30 kHz. SiC-MOSFET can operate at a switching frequency of 200 kHz, or even MHz-level [73].

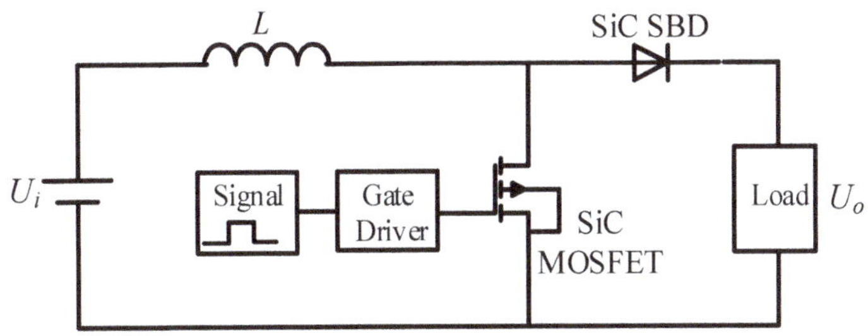

Figure 14. A topology of the boost DC–DC converter.

For Si power electronic devices, switching losses increase significantly with the increase of switching frequency. When it mentions SiC power electronic devices, the switching losses, and junction temperature are measured as shown in Figure 15, where the switching frequency of SiC-MOSFET changes from 100 kHz to 800 kHz in a 1 kW boost DC–DC converter [74]. It can be seen that both switching loss and junction temperature keep a linear relationship with switching frequency when the high-temperature and thinner layer solder is used for die attach. If the solder with a low melting point of 180 °C and low thermal conductivity is adopted for the die attach, the junction temperature can rise at an accelerating rate when the switching frequency is above 500 kHz. NASA has reported that a 100 kW DC/DC converter based SiC JFET can reach up to the operating temperature of 415 °C in 2006. In 2008, Mazumder [75] reported that the efficiency of a DC/DC converter based on SiC-JFETs can reach up to 95% at 20 °C, while the efficiency of 100 V/270 V 2 kW boost converter proposed by Kosai [76] can reach up to 90% at the temperature 200 °C, the design and performance of the boost converter were evaluated over the temperature range from 20 °C to 200 °C. The capacitance variation of the output filter is also presented in [74], reporting that a 1 kW all-SiC boost converter with the output voltage of 800V can work reliably over a switching frequency range of 100 to 800 kHz, and the steady-state working junction temperature of SiC MOSFETs has been extended to 320 °C. However, the high-frequency gate drive capability and high-temperature die-attachment technology can be the issues to develop SiC-based converter operating beyond 320 °C junction temperature.

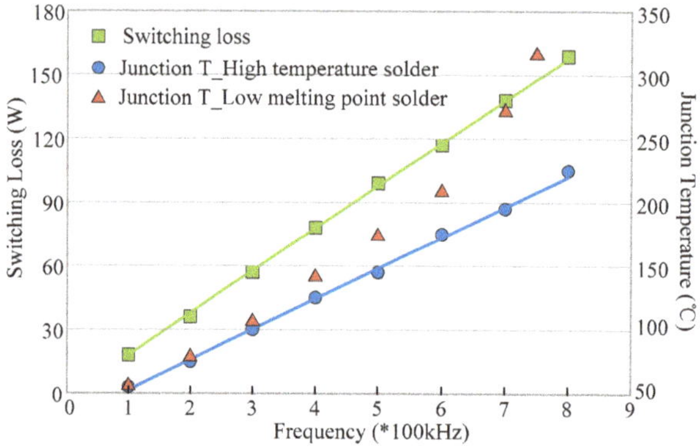

Figure 15. Switching loss and junction temperature distribution diagram under different frequencies.

3.4. SiC-based MEMS Devices

The excellent mechanical properties of SiC material, coupled with the good thermal stability at high operating temperature, offer new possibilities for developing MEMS devices for extremely harsh applications compared to those possible with Si devices.

The measurement and control technologies are also required for the high-temperature converters, and SiC devices allow the functionally integrated circuits (ICs) to operate in extreme environments. Many researchers have been working on the high-temperature ICs implemented by using SiC CMOS, JFET, and BJT, and developing the digital logic circuits, operational amplifiers, and memories. Reference [21] demonstrates that SiC makes high-temperature electronics possible up to 600 °C, reviewed the current technology performance and processing challenges relating to making ICs in SiC, and addressed that SiC devices should be commercially available in increasing quantities going forward, although the technology choice is unclear.

The first SiC-based power ICs were reported in 2008 [77]. Figure 16 shows the optical photo for 4H-SiC power integrated circuit after packaging, which includes a large power JFET and two buffer circuits. Reference [78] proposes an integrated bipolar OR/NOR gate based on 4H-SiC BJTs, and it can successfully operate up to 500 °C. References [79] and [80] report the differential amplifiers based on 4H-SiC JFET and 6H-SiC bipolar can reach up to the temperature of 500 °C and 600 °C, respectively. In [81], a 500 °C Schmitt trigger in 4H-SiC has designed and characterized, the proposed Schmitt trigger shows superior characteristics with a higher slew rate and almost independent temperature operation.

Figure 16. An optical photo for 4H-SiC power integrated circuit after packaging.

A linear voltage regulator based on the nMOS SiC has been successfully designed and tested under at 300 °C [82]. In [83], a bipolar SiC linear voltage regulator was developed to operate at 500 °C. Regarding the ICs structure, in [84], the authors propose a novel 4H-SiC lateral BJT design with symmetric and self-aligned structure, the simulation, and optimization are conducted to operate at the temperature range of 27–500 °C with an optimal current gain. is the study demonstrates that the self-aligned 4H-SiC lateral BJTs design is easier and less costly to produce, with >90% smaller than a conventional structure.

A monolithic SiC drive circuit for SiC BJTs was designed by Kargarrazi et al [85]. The performance was tested using a commercial power BJT under the resistive and capacitive conditions with the operating switching frequency up to 500 kHz. The SiC drive circuit has a good robust capability to the temperature range from 25 °C to 500 °C. In their latest publication, the controlled duty cycles from 0.5 to 0.7 are demonstrated with the operating frequency ranges from 160 to 210 kHz [86].

MEMS switches have mainly been developed in a broad swath of RF and microwave applications, and they could possibly replace positive-intrinsic-negative (PIN) diode, mechanical, FET, and other types of switches [87]. When compared to traditional micromechanical switches, MEMS switches have several advantages, such as lower insertion loss, higher isolation, and better switching figure-of-merit. They are widely used to measure oil pressure, fuel pressure and tire pressure in automotive applications, electronics, and telecom. The Foxboro is the first company who is involved in MEMS switches with the invention of the first electromechanical switch patent in the world in 1984. Analog Devices, Inc. has been started the research of MEMS switches since 1990, with the first MEMS accelerometer product successfully launched in 1991, and the first integrated MEMS gyroscope was released in 2002. The newest MEMS products released by ADI are ADGM1304 and ADGM1004, the maximum operating frequency can reach up to 14 GHz and 13 GHz, respectively. The operation temperature ranges from 0 to 85 °C, with the peak reflow soldering temperature of 260 °C. In [22], it is shown that the SiC MEMS devices are well-developed for temperatures up to 500 °C for the sensing of motion acceleration and gas flow. The digital micro switches can also be used for wireless power transfer, but the maximum operating temperature has not been reported. Reference [88] reported nanoelectromechanical system switches based inverter can operate at the temperature as high as 500 °C with ultralow leakage current, and this achievement has created a pathway toward energy-efficient high-temperature computation.

4. Challenges in High-Temperature Power Electronics

4.1. Design of High-temperature Gate Drives

Gate drives play an important role in the interface the control circuit to SiC-based devices, determining the performance of power electronics devices. Although SiC-based devices have high-speed switching capability, the drive circuit should also be matched to make full use of high-speed switching capability. Therefore, the SiC high-temperature drive circuit cannot follow the drive circuit based on conventional Si devices. Since the parasitic capacitance of a similarly-sized silicon carbide device is much lower than that of Si-based devices. A tradeoff should be taken between the component's layout and the high-speed capability, so the drive circuit should be placed at a certain distance to the power electronic devices. This will increase the gate loop and introduce larger parasitic parameters, which decrease the high-speed switching capability of the SiC-based devices in practical application. Then, the switching frequency will be limited under this condition.

The performance of SiC power electronic devices might be degraded when employed in high-temperature condition. By testing, the threshold voltage of 1.2 kV SiC MOSFET devices at 200 °C is reduced to 2/3 under normal temperature, while the on-resistance increases to 2–3 times. The reduction of threshold voltage makes the crosstalk occur more easily in the bridge-arm circuit [89]. The method of gate negative bias voltage is adapted to suppress the crosstalk, in which the negative bias voltage usually arrives at −5 V and the minimum value is −9 V. Furthermore, the active miller clamping circuit is designed to avoid the bridge-arm shoot-through. On the other hand, the gate drive board is allowed for a high-temperature environment only if the SOI die, PCB, passive components, packaging, as well as the input signal isolator, can endure the high temperature. This is expected to reduce the volume of easy chip-level integration, eliminate the high-temperature ageing effects of materials, and reduce the impact of parasitic parameters.

4.2. Current Measurement in High Temperature

The current divider and Hall sensor are usually employed to measure the current for converter control; however, they are challenging to work in the high-temperature environment. To tackle this problem, the saturated current sensor is developed for high-temperature application. While the B/H curve of magnetic materials can drift with the temperature variation, which will lead to a significant measurement error for the current sensor, a compensation algorithm or new measurement method is proposed for the high measurement accuracy. For example, the isolated DC and AC current measurement method based on a bidirectional saturated current transformer (Figure 17) can be applied to the high-temperature converter with SiC devices, which can suppress the effect of the coercive force of magnetic materials on detection precision [90].

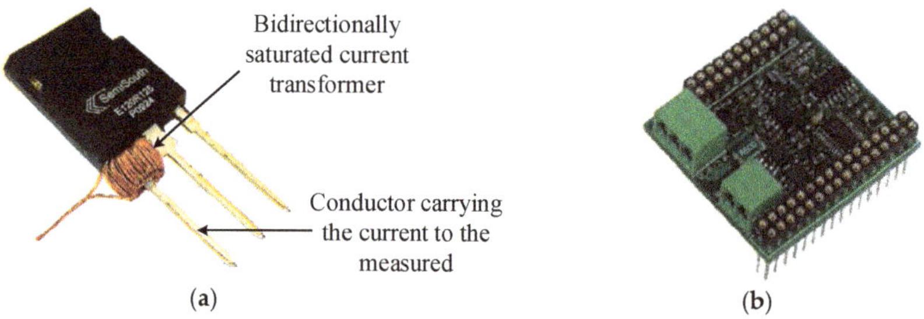

Figure 17. Fast high temperature isolated DC/AC current measurement. (**a**) Photograph of the bidirectional saturated current transformer, and (**b**) PCB circuit.

4.3. Parameters Matching within Wide Range of Temperature

The parameters matching over a wide temperature range is also an issue to design the high-temperature power electronics. The performance of the SiC components will degrade under the wide temperature cycling and the high operating temperature. With the increase of the operating temperature, the junction capacitance of SiC devices is decreased, and the switching speed is increased. The performance of passive components, such as capacitors and resistors, is also reduced, and the withstand voltage, capacitance values, and resistance values are only about 40% to 50% of normal temperature. In addition, the coefficient of thermal expansion between adjacent components should be similar; otherwise excessive mechanical stress will cause damage to the device.

Since SiC material defects and brittleness limit the size of the wafer to a small value of 4 inches in general use, the maximum current of a single chip is about 100 A. The cost of SiC chip grows exponentially with the chip current. For the application of high-power converters, the multichip power module is a cost-efficiency solution. Among paralleled chips, the parameter mismatching due to the parasitic parameters can result in the uneven electro-thermal stress. For all kinds of active and passive components, their temperature stability difference can result in the unmatched electric parameters and unbalanced mechanical stress, which significantly influence the performance and reliability of high-temperature converters.

4.4. High-temperature Packaging Technology

SiC converters and MEMS devices face the challenge of a minimum of parasitic parameter and capability of operating in a high-temperature environment. Due to the fast switching speed and low threshold voltage, SiC devices are deeply affected by the inherent and line parasitic parameters, which requires the packaging design to minimize the length of pin and wire, but the compact layout reduces the area of dissipation [91]. Likewise, the packaging technology is also a challenge to the fabrication of SiC devices for high-temperature applications. The high-temperature welding technique is one of the critical factors to improve the high-temperature capability of SiC devices [92]. The advanced material technique and the innovative structure design need to be developed to approach the SiC physical limitation.

Packaging materials and structures with improved reliability at higher temperatures are imperative for the implementation of SiC devices, mentioning the high-temperature die attach, high-conductivity TIM, and aggressive heat rejection system. For the die attach, CTE, melting temperature, porosity feature, as well as electrical and thermal conductivity are not the only indices [93]. The mechanical properties such as the modulus of elasticity, ductility, and yield strength are of equal importance [94]. Lead-free gold-based solders are a good choice for niche applications, but their mechanical stiffness is as a limiting factor, which can transfer stresses to SiC devices resulting in die cracking. The thermal resistance of TIM is still the bottleneck in most power electronics packaging, so the high-performance filler materials should be further investigated as both an enhancement and a basis for TIMs. Double-sided cooling structures, two-phase cooling methods and novel coolants can improve the cooling capability for SiC modules, which need further demonstration and understanding of long-term reliability before industrial applications.

5. Conclusions

To meet high requirements for industrial applications like the electric vehicle, electric aircraft, deep-earth oil and gas exploration, and geothermal energy development, developing the high-temperature power electronics with a significant increase of power density, efficiency, and reliability is indispensable. Compared with conventional Si power electronic devices, SiC power electronic devices have many advantages, including improved converter performance, reduced volume and weight, and simplified heat dissipation structure. This paper presented an overview of the development of high-temperature component profiles. It indicated that the advanced materials and technologies are vital for the performance promotion of high-temperature converters. Three typical application examples of SiC high-temperature converters and MEMS devices have been studied, which show the application status in several fields and uncover the main existing issues. Although many researchers are currently focusing on high-temperature packaging and integration technologies, high-temperature power electronics with SiC devices appear promising for reliable operation in a wide temperature range. Indeed, the performance of SiC converters and MEMS devices can be further enhanced when the high-temperature packaging and gate drive progress, and when the measurement and parameter matching problems are well resolved.

Author Contributions: X.G. investigated high-temperature power electronic applications, and high-temperature components, Q.X. conducted the literature reviews on chanllenges to develop high-temperature electronics, and proofread the manuscript. Z.L. collected data and references about high temperature converters, and S.D. conducted the literature reviews on high temperature MEMS devices.

Funding: This work was supported by the National Natural Science Foundation of China under grant No. 61573136, and Zhejiang Science and Technology Project under grant No. LGG18F030009 and No. 2016C31115.

Acknowledgments: Thanks to Dr. Changfu Zou for proofreading the manuscript.

Conflicts of Interest: The authors declare no conflict of interest.

References

1. Iacopi, F.; Van Hove, M.; Charles, M.; et al. Power electronics with wide band-gap materials: Toward greener, more efficient technologies. *MRS Bull.* **2015**, *40*, 390–395. [CrossRef]
2. Fraga, M.A. Comparison between the piezoresistive properties of a-SiC films obtained by PECVD and magnetron sputtering. *Silicon Carbide Relat. Mater.* **2010**, *679*, 217–220. [CrossRef]

3. Fraga, M.A.; Furlan, H.; Pessoa, R.S.; Rasia, L.A.; Mateus, C.F.R. Studies on SiC, DLC and TiO2 thin films as piezoresistive sensor materials for high temperature application. *Microsyst. Technol.* **2011**, *17*, 477–483. [CrossRef]
4. Seresinhe, R.; Lawson, C.; Madani, I. Improving the operating efficiency of the more electric aircraft concept through optimized flight procedures. *CEAS Aeronaut. J.* **2019**, *10*, 463–478. [CrossRef]
5. Phan, H.P.; Dao, D.V.; Nakamura, K.; Dimitrijev, S.; Nguyen, N. The piezoresistive effect of SiC for MEMS sensors at high temperatures: A review. *J. Microelectro Mech. Syst.* **2015**, *24*, 1663–1677. [CrossRef]
6. Balakrishnan, V.; Phan, H.P.; Dinh, T.; Dao, D.V.; Nguyen, N.-T. Thermal flow sensors for harsh environments. *Sensors* **2017**, *17*, 2061. [CrossRef] [PubMed]
7. Johnson, R.; Evans, J.; Jacobsen, P.; Thompson, J.R.; Christopher, M. The changing automotive environment: High temperature electronics. *IEEE Trans. Electron. Packag. Manuf.* **2004**, *27*, 164–176. [CrossRef]
8. Efthymiopoulos, I.; Hellier, P.; Ladommatos, N.; Russo-Profili, A.; Eveleigha, A.; Aliev, A.; Kay, A.; Mills-Lamptey, B. Influence of solvent selection and extraction temperature on yield and composition of lipids extracted from spent coffee grounds. *Ind. Crop. Prod.* **2018**, *119*, 49–56. [CrossRef]
9. Singh, D. Venus night side surface temperature. *Sci. Rep.* **2019**, *9*, 1137. [CrossRef]
10. Ning, P.; Wang, F.; Ngo, K. Forced-air cooling system design underweight constraint for high-temperature SiC converter. *IEEE Trans. Power Electron.* **2014**, *29*, 1998–2007. [CrossRef]
11. Xu, S.Z.; Peng, Y.F.; Li, S.Y. Application thermal research of forced-air cooling system in high power NPC three-level inverter based on power module block. *Case Stud. Therm. Eng.* **2016**, *8*, 387–397. [CrossRef]
12. Wu, R.; Hong, T.; Cheng, Q.; Zhou, H.; Fan, Y.; Luo, X. Thermal modeling and comparative analysis of jet impingement liquid cooling for high power electronics. *Int. J. Heat Mass Transf.* **2019**, *137*, 42–51. [CrossRef]
13. Mishra, M.K.; Dubey, V.; Mishra, P.M.; Khan, I. MEMS Technologies: A Review. *J. Eng. Res. Rep.* **2019**, *4*, 1–24.
14. Hazra, S.; De, A.; Cheng, L.; Palmour, J.; Schupbach, M.; Hull, B.A.; Allen, S.; Bhattacharya, S. High switching performance of 1700-V, 50-A SiC power MOSFET over Si IGBT/Bi MOSFET for advanced power conversion applications. *IEEE Trans. Power Electron.* **2016**, *31*, 4742–4754.
15. Garrido-Diez, D.; Baraia, L. Review of wide band-gap materials and their impact in new power devices. In Proceedings of the 2017 IEEE International Workshop of Electronics, Control, Measurement, Signals and Their Application to Mechatronics (ECMSM), Donostia-San Sebastian, Spain, 24–26 May 2017; pp. 1–6.
16. Tian, Y.; Zetterling, C.M. A fully integrated silicon-carbide sigma-delta modulator operating up to 500 °C. *IEEE Trans. Electron. Devices* **2017**, *64*, 2782–2788. [CrossRef]
17. Dreike, P.; Fleetwood, D.; King, D.; Sprauer, D.C.; Zipperian, T.E. An overview of high-temperature electronic device technologies and potential applications. *IEEE Trans. Compon. Package Manuf. Technol.* **1993**, *17*, 594–609. [CrossRef]
18. Xu, F.; Han, T.J.; Jiang, D.; Tolbert, L.M.; Wang, F.; Nagashima, J.; Kim, S.J.; Kulkarni, S. Development of a SiC JFET-based six-pack power module for a fully integrated inverter. *IEEE Trans. Power Electron.* **2013**, *28*, 1464–1478. [CrossRef]
19. Wang, R.; Chen, Z.; Boroyevich, D.; Jiang, L.; Yao, Y.; Rajashekara, K. A novel hybrid packaging structure for high-temperature SiC power modules. *IEEE Trans. Ind. Appl.* **2013**, *49*, 1609–1618. [CrossRef]
20. Khazaka, R.; Mendizabal, L.; Henry, D.; et al. Survey of high-temperature reliability of power electronics packaging components. *IEEE Trans. Power Electron.* **2015**, *30*, 2456–2464. [CrossRef]
21. Zetterling, C.M. Integrated circuits in silicon carbide for high-temperature applications. *Mater. Res. Soc. Bull.* **2015**, *40*, 431–438. [CrossRef]
22. Wright, N.G.; Horsfall, A.B. SiC sensors: A review. *J. Phys. D Appl. Phys.* **2007**, *40*, 6345–6354. [CrossRef]
23. Buttay, C.; Planson, D.; Allard, B.; Bergogne, D.; Bevilacqua, P.; Joubert, C.; Lazar, M.; Martin, C.; Morel, H.; Tournier, D.; et al. State of the art of high-temperature power electronics. *Mater. Sci. Eng. B* **2011**, *176*, 283–288. [CrossRef]
24. Watson, J.; Castro, G. A review of high-temperature electronics technology and applications. *J. Mater. Sci. Mater. Electron.* **2015**, *26*, 9226–9235. [CrossRef]
25. Horowitz, K.; Remo, T.; Reese, S. *A Manufacturing Cost and Supply Chain Analysis of SiC Power Electronics Applicable to Medium-Voltage Motor Drives. Technical Report Clean Energy Manufacturing Analysis Center;* National Renewable Energy Laboratory: Golden, CO, USA, 2017.

26. Xun, Q.; Xun, B.Y.; Li, Z.X.; Wang, P.L.; Cai, Z.D. Application of SiC power electronic devices in secondary power source for aircraft. *Renew. Sustain. Energy Rev.* **2017**, *70*, 1336–1342. [CrossRef]
27. STW100N65G2AG datasheet. Automotive-Grade Silicon Carbide Power MOSFET 650 V, 100 A, 2 0mΩ (typ., T_J = 25 °C), in a HiP247^{TM} Package. 2018. Available online: https://www.st.com/en/power-transistors/sctw100n65g2ag.html (accessed on 19 June 2019).
28. Sarnago, H.; Lucia, O.; Mediano, A.; Burdío, J.M. Improve operation of SiC-BJT based series resonant inverter with optimized base drive. *IEEE Trans. Power Electron.* **2014**, *29*, 5097–5101. [CrossRef]
29. Chen, Z.; Yao, Y.; Boroyevich, D.; Ngo, k.D.T.; Mattavelli, P.; Rajashekara, K. A 1200-V, 60-A SiC MOSFET multichip phase-leg module for high-temperature, high-frequency applications. *IEEE Trans. Power Electron.* **2014**, *29*, 2307–2320. [CrossRef]
30. Schupbach, R.M.; McPherson, B.; McNutt, T.; Lostetter, A.B.; Kajs, J.P.; Castagno, S.G. High temperature (250 °C) SiC power module for military hybrid electrical vehicle applications. In Proceedings of the NDIA Ground Vehicle Systems Engineering and Technology Symposium, Dearborn, MI, USA, 9–11 August 2011; pp. 1–7.
31. Palmour, J.W.; Kong, H.S.; Davis, R.F. Characterization of device parameters in high-temperature metal-oxide-semiconductor field-effect transistors in N-SiC thin films. *Appl. Phys.* **1988**, *64*, 2168–2177. [CrossRef]
32. Kimoto, T.; Cooper, J.A. *Fundamentals of Silicon Carbide Technology*; Wiley: New York, NY, USA, 2014.
33. Moench, S.; Costa, M.S.; Barner, A.; Weimer, J.; Holz, R.; Schwindt, E.; Zehelein, M.; Roth-Stielow, J.; Kallfass, I. High thermal conductance AlN power module with hybrid integrated gate drivers and SiC trench MOSFETs for 2kW single-phase PV inverter. In Proceedings of the 18th European Conference on Power Electronics and Applications (EPE'16 ECCE Europe), Karlsruhe, Germany, 5–9 September 2016; pp. 1–8.
34. Passmore, B.S.; Lostetter, A.B. A review of SiC power module packaging technologies: Attaches, interconnections, and advanced heat transfer. In Proceedings of the 2017 IEEE International Workshop on Integrated Power Packaging (IWIPP), Delft, The Netherlands, 5–7 April 2017; pp. 1–5.
35. Cook, G.O.; Sorensen, C.D. Overview of transient liquid phase and partial transient liquid phase bonding. *J. Mater. Sci.* **2011**, *46*, 5305–5323. [CrossRef]
36. Shen, Z.Z.; Johnson, R.W.; Hamilton, M.C. SiC power device die attach for extreme environments. *IEEE Trans. Electron. Devices* **2015**, *62*, 346–353. [CrossRef]
37. Menon, S.; George, E.; Osterman, M.; Pecht, M. High lead solder (Over 85%) solder in the electronics industry: RoHS exemptions and alternatives. *J. Mater. Sci. Mater. Electron.* **2015**, *26*, 4021–4030. [CrossRef]
38. Mallampati, S.; Yin, L.; Shaddock, D.; Schoeller, H.; Cho, J. Lead-free alternatives for interconnects in high-temperature electronics. *Asme J. Electron. Packag.* **2018**, *140*, 010906. [CrossRef]
39. Wintrich, A.; Nicolai, U.; Tursky, W.; Reimann, T. *Application Manual Power Semiconductors*, 2nd ed.; SEMIKRON International GmbH: Nuremberg, Germany, 2015.
40. Zhang, T.; Sammakia, B.G.; Yang, Z.; Wang, H. Hybrid nanocomposite thermal interface materials: The thermal conductivity and the packing density. *Asme J. Electron. Packag.* **2018**, *140*, 031006. [CrossRef]
41. Peacock, M.A.; Roy, C.K.; Hamilton, M.C.; Johnson, R.W.; Knight, R.W.; Harris, D.K. Characterization of transferred vertically aligned carbon nanotubes arrays as thermal interface materials. *Int. J. Heatmass Transf.* **2016**, *97*, 94–100. [CrossRef]
42. Broughton, J.; Smet, V.; Tummala, R.R.; Joshi, Y.K. Review of thermal packaging technologies for automotive power electronics for traction purposes. *J. Electron. Packag.* **2018**, *140*, 40801. [CrossRef]
43. Liang, G.; Mudawar, I. Review of spray cooling—Part 1: Single-phase and nucleate boiling regimes, and critical heat flux. *Int. J. Heat Mass Transf.* **2017**, *115*, 1174–1205. [CrossRef]
44. Huang, H.; Borhani, N.; Richard Thome, J. Thermal response of multi-microchannelevaporators during flow boiling of refrigerants under transient heat loads with flow visualization. *Asme J. Electron. Packag.* **2016**, *138*, 031004. [CrossRef]
45. Lee, H.; Agonafer, D.D.; Won, Y.; Houshmand, F.; Gorle, C.; Asheghi, M.; Goodson, K.E. Thermal modeling of extreme heat flux micro channel coolers for GaN-on-SiC semiconductor devices. *Asme J. Electron. Packag.* **2016**, *138*, 010907. [CrossRef]
46. Huque, M.A.; Tolbert, L.M.; Blalock, B.J.; Islam, S.K. Silicon-on-insulator-based high-voltage, high-temperature integrated circuit gate driver for silicon carbide-based power field effect transistors. *IET Power Electron.* **2010**, *3*, 1001–1009. [CrossRef]

47. Qi, F.; Xu, L.Y. Development of a High-Temperature Gate Drive and Protection Circuit Using Discrete Components. *IEEE Trans. Power Electron.* **2017**, *32*, 2957–2963. [CrossRef]
48. Huque, M.A.; Islam, S.K.; Tolbert, L.M.; Blalock, B.J. A 200 °C universal gate driver integrated circuit for extreme environment applications. *IEEE Trans. Power Electron.* **2012**, *27*, 4153–4162. [CrossRef]
49. Wang, Z.Q.; Shi, X.J.; Tolbert, L.M.; Wang, F.; Liang, Z.; Costinett, D.; Blalock, B.J. A high temperature silicon carbide MOSFET power module with integrated silicon-on-insulator-based gate drive. *IEEE Trans. Power Electron.* **2015**, *30*, 1432–1445. [CrossRef]
50. Yao, Y.Y.; Lu, G.Q.; Boroyevich, D.; Ngo, K.D.T. Survey of high-temperature polymeric encapsulants for power electronics packaging. *IEEE Trans. Compon. Packag. Manuf. Technol.* **2015**, *5*, 168–181.
51. Ma, B.; Tae, H.; Stephen, L.; Dorris, E.; Koritala, R.E.; Balachandran, U. Flexible ceramic film capacitors for high-temperature power electronics. *Mater. Sci. Energy Technol.* **2019**, *2*, 96–103. [CrossRef]
52. Ericson, N.; Frank, S.; Britton, C.; Marlino, L.; Ryu, S.-H.; Grider, D.; Mantooth, A.; Francis, M.; Lamichhane, R.; Mudholkar, M.; et al. A 4H silicon carbide gate buffer for integrated power systems. *IEEE Trans. Power Electron.* **2014**, *29*, 539–542. [CrossRef]
53. Alexandru, M.; Banu, V.; Jordà, X.; Montserrat, J.; Vellvehi, M.; Tournier, D.; Millán, J.; Godignon, P. SiC integrated circuit control electronics for high-temperature operation. *IEEE Trans. Ind. Electron.* **2015**, *62*, 3182–3191. [CrossRef]
54. Ghandi, R.; Chen, C.P.; Yin, L.; Zhu, X.; Yu, L.; Arthur, S.; Ahmad, F.; Sandvik, P. silicon carbide integrated circuits with stable operation over a wide temperature range. *IEEE Trans. Electron. Device* **2014**, *35*, 1206–1208. [CrossRef]
55. Peftitsis, D.; Rabkowski, J. Gate and base drivers for silicon carbide power transistors: An overview. *IEEE Trans. Power Electron.* **2016**, *31*, 7194–7213. [CrossRef]
56. Nayak, P.; Pramanick, S.K.; Rajashekara, K. A high-temperature gate driver for silicon carbide MOSFET. *IEEE Trans. Ind. Electron.* **2018**, *65*, 1955–1964. [CrossRef]
57. Wang, R.X.; Boroyevich, D.; Ning, P.Q.; Wang, Z.Q.; Wang, F.; Mattavelli, P.; Ngo, K.D.T.; Rajashekara, K. A high temperature SiC three-phase ac-dc converter design for >100 °C ambient temperature. *IEEE Trans. Power Electron.* **2013**, *28*, 555–572. [CrossRef]
58. Li, Q.; Chen, L.; Gadinsk, M.R.; Zhang, S.; Zhang, G.; Li, H.U.; Iagodkine, E.; Haque, A.; Chen, L.Q.; Jackson, T.N.; et al. Flexible high-temperature dielectric materials from polymer nanocomposites. *Nature* **2015**, *523*, 576–579. [CrossRef]
59. Reynolds, C. Tantalum capacitor technology: Options for high0temperature and harsh-environment applications. *IEEE Power Electron. Mag.* **2017**, *4*, 43–47. [CrossRef]
60. Bunel, C.; Lengignon, L. Silicon capacitors with extremely high stability and reliability ideal for high temperature applications. *Addit. Conf.* **2012**, *2012*, 273–277. [CrossRef]
61. Bartnitzek, T.; Thelemann, T.; Apel, S.; Suphan, K.H. Advantages and limitations of ceramic packaging technologies in harsh applications. *Int. Symp. Microelectron.* **2016**, *1*, 581–585. [CrossRef]
62. Hessien, M.M.; Rashad, M.M.; El-Barawy, K.; Ibrahim, I.A. Influence of manganese substitution and annealing temperature on the formation, microstructure and magnetic properties of Mn-Zn ferrites. *J. Magn. Magn. Mater.* **2008**, *320*, 1615–1621. [CrossRef]
63. Cooke, M. Wide load potential for electric vehicles. *Compd. Adv. Silicon* **2009**, *4*, 70–75.
64. Qin, H.; Ma, C.; Zhu, Z.; Yan, Y. Influence of Parasitic Parameters on Switching Characteristics and Layout Design Considerations of SiC MOSFETs. *J. Power Electron.* **2018**, *18*, 1255–1267.
65. Zhang, H.; Tolbert, L.M.; Han, J.H.; Chinthavali, M.S.; Barlow, F. 18 kW three phase inverter system using hermetically sealed SiC phase-leg power modules. In Proceedings of the twenty-fifth annual IEEE applied power electronics conference and exposition (APEC), Palm Springs, CA, USA, 21–25 February 2010; pp. 1108–1112.
66. APEI Company. Multichip Power Module Technology. Available online: https://arpa-e.energy.gov/?q=impact-sheet/apei-adept (accessed on 19 June 2019).
67. Cao, W.; Mecrow Barrie, C.; Glynn, J.; Bennett, J.W.; Atkinson, D.J. Overview of electric motor technologies used for more electric aircraft (MEA). *IEEE Trans. Ind. Electron.* **2012**, *59*, 3523–3531.
68. Wada, K.J.; Ando, M. Limitation of DC-side stray inductance by considering over voltage and short-circuit current. In Proceedings of the 2013 15th European Conference on Power Electronics and Applications (EPE), Lille, France, 2–6 September 2013; pp. 1–8.

69. Li, H.; Munk-Nielsen, S.; Wang, X.; Maheshwari, R.; Bęczkowski, S.; Uhrenfeldt, C.; Franke, W.T. Influences of device and circuit mismatches on paralleling silicon carbide MOSFETs. *IEEE Trans. Power Electron.* **2016**, *31*, 621–634. [CrossRef]
70. Ning, P.Q.; Lei, T.G.; Wang, F.; Lu, G.Q.; Ngo, K.D.T.; Rajashekara, K. A novel high-temperature planar package for SiC multichip phase-leg power module. *IEEE Trans. Power Electron.* **2010**, *25*, 2059–2067. [CrossRef]
71. Zhang, D.; Ning, P.; Boroyevich, D. Development of an all SiC high-power density three-phase rectifier with interleaving. In Proceedings of the IEEE Energy Conversion Congress and Exposition, Phoenix, AZ, USA, 17–22 September 2011; pp. 4073–4080.
72. Sun, P.W.; Lai, J.S.; Liu, C.; Yu, W. A 55-kW three-phase inverter based on hybrid-switch soft-switching modules for high-temperature hybrid electric vehicle drive application. *IEEE Trans. Ind. Appl.* **2012**, *48*, 962–969. [CrossRef]
73. Guo, S.; Zhang, L.; Lei, Y.; Li, X.; Xue, F.; Yu, W.; Huang, A.Q. 3.38MHz operation of 1.2kV SiC MOSFET with integrated ultra-fast gate drive. In Proceedings of the IEEE 3rd Workshop on Wide Band-Gap Power Devices and Applications (WiPDA), Blacksburg, VA, USA, 2–4 November 2015; pp. 390–395.
74. Zhong, X.Q.; Wu, X.K.; Zhou, W.C.; Sheng, K. An all-SiC high-frequency boost DC–DC converter operating at 320 °C junction temperature. *IEEE Trans. Power Electron.* **2014**, *29*, 5091–5096. [CrossRef]
75. Mazumder, S.K.; Acharya, K.; Jedraszczak, P. High-temperature all-SiC bidirectional DC/DC converter for plug-in-hybrid vehicle (PHEV). In Proceedings of the 34th Annual Conference of IEEE Industrial Electronics, Orlando, FL, USA, 10–13 November 2008; pp. 2873–2878.
76. Kosai, H.; Scofield, J.; McNeal, S.; Jordan, B.; Ray, B. Design and performance evaluation of a 200 °C interleaved boost converter. *IEEE Trans. Power Electron.* **2013**, *28*, 1691–1699. [CrossRef]
77. Sheng, K.; Zhang, Y.; Su, M.; Zhao, J.H.; Li, X.; Alexandrov, P.; Fursin, L. Demonstration of the first SiC power integrated circuit. *Solid State Electron.* **2008**, *52*, 1636–1646. [CrossRef]
78. Lanni, L.; Ghandi, R.; Zetterling, C.M.; Malm, B.G.; Östling, M. 500 °C bipolar integrated OR/NOR gate in 4H-SiC. *IEEE Electron. Device Lett.* **2013**, *34*, 1091–1093. [CrossRef]
79. Hedayati, R.; Lanni, L.; Rodriguez, S.; Malm, B.G.; Rusu, A.; Zetterling, C.-M. A monolithic, 500 °C operational amplifiers in 4H-SiC bipolar technology. *IEEE Electron. Device Lett.* **2014**, *35*, 693–695.
80. Patil, A.C.; Fu, X.; Mehregany, M.; Garverick, S.L. Fully-monolithic, 600 °C differential amplifiers in 6H-SiC JFET IC technology. In Proceedings of the IEEE Custom Integrated Circuits Conference, Rome, Italy, 13–16 September 2009; pp. 73–76.
81. Kargarrazi, S.; Lanni, L.; Zetterling, C. Design and characterization of 500 °C Schmitt Trigger in 4H-SiC. *Mater. Sci. Forum* **2014**, *821*, 897–901. [CrossRef]
82. JValle-Mayorga, A.; Rahman, A.; Mantooth, H.A. A siC NMOS linear voltage regulator for high-temperature applications. *IEEE Trans. Power Electron.* **2014**, *29*, 2321–2328. [CrossRef]
83. Kargarrazi, S.; Lanni, L.; Saggini, S.; Rusu, A.; Zetterling, C.-M. 500 °C bipolar SiC linear voltage regulator. *IEEE Trans. Electron. Devices* **2015**, *62*, 1–5. [CrossRef]
84. Siddiqui, A.; Elgabra, H.; Singh, S. Design considerations for 4H-SiC lateral BJTs for high temperature logic applications. *J. Electron. Devices Soc.* **2018**, *6*, 126–134. [CrossRef]
85. Kargarrazi, S.; Lanni, L.; Rusu, A.; Zetterling, C.M. A monolithic SiC drive circuit for SiC power BJTs. In Proceedings of the International Symposium on Power Semiconductor Devices & IC's, Hong Kong, China, 10–14 May 2015; pp. 285–288.
86. Kargarrazi, S.; Elahipanah, H.; Asggini, S.; Senesky, D.; Zetterling, C.M. 500 °C SiC PWM integrated circuit. *IEEE Trans. Power Electron.* **2019**, *34*, 1997–2001. [CrossRef]
87. Basu, A.; Adams, G.G.; McGruer, N.E. A review of micro-contact physics, materials, and failure mechanisms in direct-contact RF MEMS switches. *J. Micromech. Microeng.* **2016**, *26*, 104004. [CrossRef]
88. Lee, T.; Bhunia, S.; Mehregany, M. Electromechanical computing at 500 °C with silicon carbide. *Science* **2010**, *329*, 1316–1318. [CrossRef] [PubMed]
89. Yin, S.; Tseng, K.J.; Tong, C.F.; Simanjorang, R.; Gajanayake, C.J.; Nawawi, A.; Liu, Y.; Liu, Y.; See, K.Y.; Sakanova, A.; et al. Gate driver optimization to mitigate shoot-through in high-speed switching SiC half bridge module. In Proceedings of the IEEE International Conference on Power Electronics & Drive Systems, Sydney, Australia, 9–12 June 2015; pp. 484–491.

90. Wrzecionko, B.; Steinmann, L.; Kolar, J.W. High-bandwidth high-temperature (250 °C/500 °F) isolated DC and AC current measurement: Bi-directionally saturated current transformer. *IEEE Trans. Power Electron.* **2013**, *28*, 5404–5413. [CrossRef]
91. Trentin, A.; Zanchetta, P.; Wheeler, P.; Clare, J. Performance evaluation of 3 phase buck-type PWM rectifiers with integrated and symmetrical boost converter using SiC MOSFETs for aircraft application. In Proceedings of the IEEE Energy Conversion Congress and Exposition, Denver, CO, USA, 15–19 September 2013; pp. 1–6.
92. Qi, F.; Xu, L.; Zhao, G.; Wang, J. Transformer isolated gate drive with protection for SiC MOSFET in high temperature application. In Proceedings of the IEEE in Energy Conversion Congress and Exposition, Pittsburgh, PA, USA, 14–18 September 2014; pp. 5723–5728.
93. Siow, K.S. Are Sintered Silver Joints Ready for Use as Interconnect Material in Microelectronic Packaging? *J. Electron. Mater.* **2014**, *43*, 947–961. [CrossRef]
94. Manikam, V.R.; Cheong, K.Y. Die attach materials for high temperature applications: A review. *IEEE Trans. Compon. Packag. Manuf. Technol.* **2011**, *1*, 457–478. [CrossRef]

micromachines

MDPI

Article

Advances in High-Resolution Radiation Detection Using 4H-SiC Epitaxial Layer Devices

Krishna C. Mandal *, Joshua W. Kleppinger and Sandeep K. Chaudhuri

Department of Electrical Engineering, University of South Carolina, Columbia, SC 29208, USA; jwk@email.sc.edu (J.W.K.); chaudhsk@mailbox.sc.edu (S.K.C.)

* Correspondence: mandalk@cec.sc.edu; Tel.: +1-803-777-2722

Received: 29 January 2020; Accepted: 25 February 2020; Published: 28 February 2020

Abstract: Advances towards achieving the goal of miniature 4H-SiC based radiation detectors for harsh environment application have been studied extensively and reviewed in this article. The miniaturized devices were developed at the University of South Carolina (UofSC) on 8 × 8 mm 4H-SiC epitaxial layer wafers with an active area of ≈11 mm^2. The thicknesses of the actual epitaxial layers were either 20 or 50 μm. The article reviews the investigation of defect levels in 4H-SiC epilayers and radiation detection properties of Schottky barrier devices (SBDs) fabricated in our laboratories at UofSC. Our studies led to the development of miniature SBDs with superior quality radiation detectors with highest reported energy resolution for alpha particles. The primary findings of this article shed light on defect identification in 4H-SiC epilayers and their correlation with the radiation detection properties.

Keywords: silicon carbide; 4H-SiC, epitaxial layer; Schottky barrier; radiation detector; point defects; deep level transient spectroscopy (DLTS); thermally stimulated current spectroscopy (TSC); electron beam induced current spectroscopy (EBIC); pulse height spectroscopy (PHS)

1. Introduction

Development of advanced high-resolution radiation detectors is aimed at those that are capable of reliable and long-term non-degrading operation at elevated temperatures under high doses of ionizing radiation. The application of such radiation detectors is primarily in the field of, but not restricted to, biomedical and nuclear engineering, medical imaging devices, and homeland security including nuclear materials accounting and safeguarding in harsh environments. Hence, to meet the criteria of field deployment, such detectors need to be compact and miniaturized. Applicability of high-resolution germanium (Ge) or silicon (Si) based detectors are extremely limited in these fields owing to their cryogenic operating temperature (low bandgap energy) and lack of radiation hardness. Room temperature compact detectors like cadmium telluride (CdTe) or cadmium zinc telluride (CdZnTe, CZT) falls behind when it comes to high temperature applicability as these materials cannot survive high temperatures. Silicon carbide (SiC), a wide band-gap semiconductor, has emerged as a potential alternative to more mature technologies in intense and rugged environments. Due to its wide bandgap (3.27 eV at 300 K), 4H-SiC devices show extremely low reverse leakage current at room temperature and elevated temperatures as well [1–3]. Detectors based on 4H-SiC epitaxial layers with low defect densities and impurities can be used to reliably detect any type of ionizing radiation at high radiation load (dose level of 22 MGy and higher [4,5]), in conditions of aggressive medium and at elevated temperatures, in systems for monitoring in acid/alkaline-containing media, and in systems that are designed for accurately determining the field of ionizing particles, X- and γ-ray radiation. SiC does not melt under laboratory conditions but sublimes at temperatures as high as 2700 °C and is highly chemically inert. Polytype 4H-SiC has a very high threshold displacement energy (22–35 eV) which accounts for its extreme radiation hardness [6,7]. Over the past decade, the quality of SiC material has improved significantly, which added momentum to the development of SiC based devices. Recent availability of high quality 4H-SiC epitaxial layers has oriented the radiation detector community to use such epilayers for radiation detection measurements with highly promising results. Although the resolution of the detectors based on bulk semi-insulating (SI) SiC grown by physical vapor transport (PVT) is not adequate, presumably due to high density of defects and deep level centers [7], the Schottky barrier device (SBD) detectors fabricated using SiC epitaxial layers [8–10] perform extremely well in high-resolution detection of low penetration depth radiation. Additionally, the response of these kind of detectors for soft X-rays are significantly higher than that of commercial off-the-shelf (COTS) SiC UV photodiode.

The present article covers our recent results and reviews the advances towards the goal of achieving high-resolution and compact radiation detectors to be used in harsh environmental conditions. The majority of the developments in this article were carried out at Photovoltaics and Nuclear Radiation Accounting Devices (PANRAD) laboratory at the University of South Carolina (UofSC). Devices on 4H-SiC epitaxial layers were fabricated solely in our laboratory and characterized in terms of physical properties, epitaxial layer quality, electrical and radiation detection properties, and defect levels' identification and quantification. The article focusses on the investigation of electrical properties like electron-hole pair creation energy and minority carrier diffusion length, and device properties such as surface barrier height, built-in potential, and leakage current. Point and structural defects that limit device performance were discussed in detail in the various types of epitaxial layers through defect delineating etching and XRD rocking curve measurements, electron beam induced current spectroscopy (EBIC), thermally stimulated current (TSC) spectroscopy, and deep level transient spectroscopy (DLTS). Electronic noise characteristics of the associated front-end electronics which affect the radiation detection properties of 4H-SiC for high-energy alpha particles were also investigated and presented. We also cover pulse height spectra (PHS) with 5486 keV alpha particles and 59.6 keV γ-rays from radiation sources, and soft X-ray responsivity measurements performed at the National Synchrotron Light Source (NSLS) at Brookhaven National Laboratory (BNL).

2. Materials and Methods

In this article we present the results obtained from 20 and 50 µm thick n-type epitaxial layer grown on 76 mm diameter 4H-SiC (0001) wafer, which was highly doped with nitrogen and off-cut 8° towards the $[11\bar{2}0]$ direction. The typical effective doping concentration in these epitaxial layers measured using high frequency (100 kHz) capacitance-voltage (C-V) method was found to be 1.8×10^{14}–3.2×10^{14} cm^{-3}. The radiation detectors were fabricated on 8×8 mm^2 substrates diced from the 76 mm diameter wafer by depositing 3.2–3.8 mm in diameter and $\approx$10 nm in thickness Ni Schottky contacts on top of the epitaxial layers through a shadow mask and using a Quorum model Q150T sputtering unit. Large Ni contacts ($\approx 6 \times 6$ mm^2) of $\approx$100 nm in thickness were deposited on the other side of the wafer by the same means. A schematic of a typical cross-sectional view of the epitaxial layer with nickel contacts structure is shown in Figure 1.

For the semi-insulating (SI) 4H-SiC detector fabrication, $\approx$390 µm thick 8×8 mm^2 substrates, diced from (0001) 76 mm semi-insulating wafer with resistivity $\geq 10^{12}$ Ω-cm, were used. The electrical contacts on the test structures were achieved by electron-beam deposition of 3.2 and 1.2 mm diameter Ni contacts on the Si face of the substrate and 3.2 mm in diameter or larger size ($\approx 6 \times 6$ mm^2) Ni contact onto the C face of the substrate using appropriately designed shadow masks. The substrates were cleaned using a standard Radio Corporation of America (RCA) cleaning procedure [11] prior to contact deposition. No annealing treatment was conducted after the deposition of Ni contacts unless otherwise mentioned in this article. For the high performing samples, the RCA cleaning of the wafers and the electrical contact deposition were achieved using photolithography techniques performed in class 100 clean room facilities. It may be noted here that the thickness of the Ni contact plays a pivotal role in defining the energy resolution of the detector. Thicker contacts may add to the variation of the incident energy due to excessive scattering. We deposited Ni contacts as thin as possible and simultaneously inspected whether the contact produced reliable electrical contacts or not. We adopted the minimum thickness as the optimized thickness of detector window, which produced reliable contacts confirmed from repeated I-V measurements.

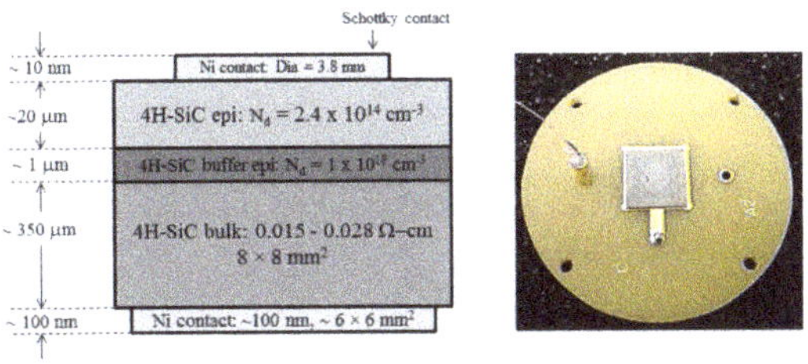

Figure 1. A cross-sectional schematic view of the epitaxial layer structure and a single pixel n-type Ni/4H-SiC Schottky barrier device (SBD) mounted on a printed circuit board (PCB). The square shaped back contact is visible through the wafer. The circular contact on the top is connected using a 25 μm thin gold wire.

A wire bonding technique was developed in our laboratory for achieving strong and durable electrical connections with the Ni contacts and minimizing the bonding area for efficient radiation detection. This technique involves special type of silver epoxy rated for high temperature and vacuum applications. The same type of the silver epoxy was used for mounting the chip on a printed circuit board (PCB). A photograph of a typical single pixel detector is shown in Figure 1.

TSC measurements on the epitaxial layer were conducted in the temperature range 94–550 K in vacuum $< 1 \times 10^{-4}$ Torr at 4–15 K/min heat rates. The trap filling was achieved by illuminating the samples at 94 K using UVP model UVM-57 Handheld UV Lamp specified to produce 302 nm UV light. The detailed description of our TSC measurement set-up is available in our earlier work [12].

Current-voltage (I-V) characterizations were performed using a Keithley (Cleveland, OR, USA) 237 high voltage source-measure unit. For the low temperature measurements, a vacuum cryo-chamber from MMR Technologies (CA, USA) was used. Capacitance-voltage (C-V) measurements were performed using either a Keithley (Cleveland, OR, USA) 590 CV analyzer or the DLTS measurement unit.

The DLTS measurements were carried out using a SULA (Ashland, OR, USA) DDS-12 modular DLTS system. The DLTS system comprised of a pulse generator module for applying repetitive bias pulses, a 1 MHz oscillator for capacitance measurements, a sensitive capacitance-meter involving self-balancing bridge circuit, and a correlator/pre-amplifier module which automatically removes DC background from the capacitance meter and amplifies the resultant signal change. The correlators were based on a modified double boxcar signal averaging system. The sample was mounted in a Janis (Woburn, MA, USA) VPF 800 LN2 cryostat for temperature variation, which was controlled by a Lakeshore (Cleveland, OH, USA) LS335 temperature controller. The DDS-12 system allows the user to collect four DLTS spectra simultaneously corresponding to four different rate windows in a single temperature scan. The signals were digitized using a national instruments (NI, Austin, TX, USA) digitizer card integrated with the DLTS system for on line processing using a personal computer (PC). The entire system including the modules and the temperature controller is controlled using a dedicated LabVIEW interface, which also allows the user to analyze the recorded data.

In order to evaluate the density of crystallographic defects, defect delineating chemical etching in molten KOH was performed at ≈825 K for about 5 min. Threading edge, screw, and basal plane dislocation densities (BPDs) were assessed via etch pit density (EPD) measurements using a Nomarski optical microscope (Make and model). X-ray diffraction rocking curves were acquired using a double crystal diffractometer (DSO-1), by Radicon Scientific Instruments Co. (Pune, India). We used CuK_α radiation and (0008) reflection in the rocking curve measurements. The EBIC measurements were carried out at 29 kV accelerating voltage and 0 V bias voltage using a JEOL (Peabody, MA, USA) 35 SEM scanning electron microscope (SEM).

A vacuum annealing apparatus was specially designed in our laboratory for the isochronal annealing studies. The annealing set-up comprised of a furnace with a temperature control unit, an oil-free mechanical pump, a turbo-molecular pump, vacuum gauge controllers, and turbo pump controller. The samples were loaded in quartz ampoules and placed under vacuum on the order of 2×10^{-7} torr. Upon reaching the desired temperature, the sample under vacuum was lowered to the hot zone of a single-zone tubular furnace. The samples were annealed for a duration of $\approx$30 min. Before each annealing stage, the top and bottom nickel contact of the Schottky devices were completely etched-off using concentrated nitric acid (HNO_3). After annealing at each temperature, the samples were cleaned by RCA cleaning procedure and Ni Schottky and Ohmic contacts were deposited for subsequent I-V, C-V, and DLTS measurements.

Pulse height spectroscopy (PHS) was carried out using a spectrometer comprising of a pre-amplifier (Cremat, Newton, MA, USA), CR-110 or Amptek (Bedford, MA, USA) A250 CoolFET) whose signals were fed to an Ortec (Oak Ridge, TN, USA) 572 spectroscopy amplifier. The shaped signals were digitized and binned to obtain pulse-height spectra using a Canberra (Yvelines, France) Multiport II multichannel analyzer (MCA) unit, controlled by Genie 2000 (Canberra (Yvelines, France)) interface software. Pulse-height spectra of 5486 keV alpha particles and 59.6 keV gamma rays were obtained using a $\approx$1.0 μCi ^{241}Am (alpha and X-γ ray) radiation source. The source and the detector were placed inside an electromagnetic interference shielded aluminum box, which was constantly evacuated during the data acquisition using a vacuum pump, in order to minimize scattering of alpha particles with air molecules.

The detectors fabricated in our laboratory at UofSC using n-type 4H-SiC epitaxial layers were also tested and evaluated at Los Alamos National Laboratory (LANL, Los Alamos, NM, USA) for detecting low energy X-rays and compared to commercial-off-the-self (COTS) SiC UV photodiode detectors. The measurements were performed at 20-250 V bias voltages using U3C and X8A beam lines [13] at the National Synchrotron Light Sources (NSLS) at Brookhaven National Laboratory (BNL, New York, NY, USA). This beam line provides monochromatic photon beams ranging from 50 to 6500 eV with intensities as high as 10^{12} photons/second.

3. Results

The following results were obtained using various 4H-SiC epitaxial layer detectors fabricated at UofSC. The wafers were categorized based on their performance as radiation detectors and are tabulated below in Table 1.

Table 1. Categorization of 4H-SiC wafers used in this study.

Conductivity	Thickness (μm)	Quality	Designation
Semi-insulating	50	Good	SI-M50
n-typen	50	Good	n-M50
n-typen	50	Superior	n-S50
n-typen	20	Superior	n-S20

3.1. *Electrical Measurements on SI and n-Type Epitaxial Layers*

Figure 2 represents the I-V characteristics for semi-insulating 4H-SiC epitaxial (SI-M50) samples [14]. It can be observed from the graphs that the I-V characteristics are asymmetric with respect to the polarity of the applied bias. Currents higher than 10 μA at room temperature were observed for positive bias as low as 20 V on the Ni/SI-SiC surface. As a result, a positive bias will be referred to as the forward bias henceforth. The asymmetry of the I-V curves are thought to be due to the different potential distribution at the Ni/SI-SiC and SI-SiC/n$^+$SiC interfaces [14]. It could also be observed from Figure 2 that each I-V curve at a given temperature in the forward bias region consists of two branches corresponding to ramping the bias voltage up and down showing a hysteresis. The difference is pronounced for temperatures below 600 K. Further, the reverse I-V characteristics show step like features. Step 1 with $V_{S1} \approx -70V$ was observed at temperatures above 200 K and step 2 with $V_{S2} \approx -1V$ occurred at temperatures above 350 K. The steps in the I-V characteristics normally correspond to the ultimate filling of trap centers by injected carriers [15,16]. The voltage V_s at which the step onsets, is given by,

$$V_s = \frac{qL^2}{2\varepsilon_{SiC}} \times p_{t,0} \tag{1}$$

The concentration of trap centers with activation energy E_t, unoccupied by the injected carriers, $(p_{t,0})$ is given by $\frac{N_t}{g}\exp\left[\frac{E_t-F_0}{k_BT}\right]$. Here, k_B is the Boltzmann's constant, L is the thickness of the epitaxial layer, q is the electronic charge, ε_{SiC} is the permittivity of the material, F_0 is the Fermi energy, N_t is the trap concentration, and g is the degeneracy factor. Using Equation (1) for the step voltages $V_{S1} = -70$ V and $V_{S2} = -1$ V, the trap densities for the centers unoccupied by the injection carriers was calculated to be 2×10^{14} cm^{-3} and 3×10^{12} cm^{-3}, respectively. The former trap concentration was assumed to be the total trap concentration considering traps corresponding to step 1 were emptied at $V = 0$.

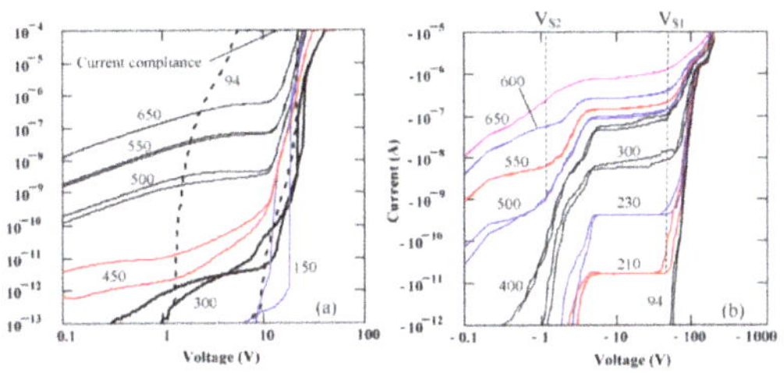

Figure 2. Forward (**a**) and reverse (**b**) I-V characteristics obtained for SI 4H-SiC sample (SI-M50) with 1 mm diameter Ni contact at various temperatures.

From a similar study on a n-type (n-M50) sample, the variation of barrier height as a function of temperature was investigated [17]. It was observed that Ni on 4H-SiC epitaxial layer formed a typical Schottky junction. Unlike the n-SI50 samples, no step like features were noticed in the reverse bias characteristics of these samples. Using the activation energy method described in [16], the electrically active area A and the effective Richardson constant A^{**} was calculated. The expression of a forward current I_F as a function of forward bias voltage V_F, in a Schottky junction can be expressed as in Equation (2).

$$\ln\left(\frac{I_F}{T^2}\right) = \ln(AA^{**}) - q(\phi_B - V_F)/k_BT \tag{2}$$

where, T is the absolute temperature, q is the electronic charge, and ϕ_B is the Schottky barrier height. Figure 3 shows the plot of $\frac{I_F}{T^2}$ as a function of inverse of absolute temperature (1000/T). A linear fit to the curve gives the the activation energy $q(\phi_B - V_F)$ and the product AA^{**} from the slope and intercepts, respectively. The studies were conducted in the temperature range 325–500 K with $V_F = 0.8$ V. Using a geometrical area of 0.08 cm^2, the effective Richardson constant was found to be 0.074 Amp/K^2.cm^2 which is three orders of magnitude lower than the theoretically predicted value of 146 Amp/K^2.cm^2. Moreover, if the theoretical value of A^{**} is considered then the effective geometrical area of the detector turns out to be 0.008 cm^2 which is an order of magnitude less than the actual geometrical area. The reason behind obtaining a lower value of A^{**} is thought to be due to using the geometrical area in the calculations, instead of the effective electrically active area. It is a well known fact that the measured currents in an I-V measurement correspond to the currents flowing through the patches with lower surface barrier height only [18,19].

The effect can be more pronounced in Schottky contacts with large spatial variation of barrier heights. As a result, the resulting effective area through which any current could flow is lower than the actual geometric contact area.

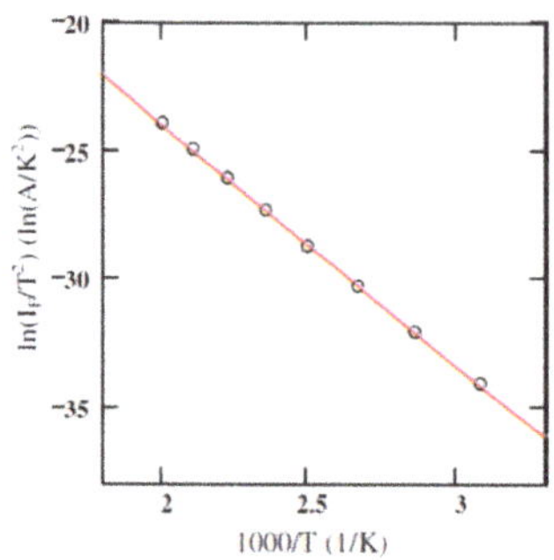

Figure 3. Activation energy plot for a 50 µm n-type 4H-SiC epitaxial layer (n-M50) SBD with 3.2 mm diameter Ni contact. The solid red line shows the linear fit to the experimental data.

Figure 4 shows the I-V characteristics obtained for a 50 µm n-type Ni/4H-SiC (n-S50) epitaxial layer detector [20]. Like the n-M50 samples, these samples also showed clearly the formation of Schottky type junction. Currents as low as 0.2 nA for −100 V was obtained and the Schottky junction was observed to be conducting for bias voltages below +1.0 V. Hence a positive bias on the Ni/n-4H-SiC interface constitutes a forward biased Schottky junction. These reverse biased I-V curves, in contrast to the SI epitaxial layers, did not exhibit any step like regions either, indicating the low concentration of those trap centers responsible for the step like structures in the SI samples. The diode ideality factor and Schottky barrier height were calculated from the forward I-V characteristics using a thermionic emission model [19]. According to the model, the current (I) for an asymmetric junction (Schottky in this case) is related to the applied bias V_a across the junction, in the following manner

$$\ln(I) \sim \left(\frac{\beta}{n}\right)\ln(V_a) + \ln(I_s) \tag{3}$$

where, I_s is the saturation current, n is the diode ideality factor, and $\beta = q/k_BT$, q being the electronic charge and T the absolute temperature. The saturation current, according to the model, is given by

$$I_s = A^{**}AT^2\exp(-\beta\phi_B) \tag{4}$$

where, A^{**} is the effective Richardson constant (146 $Acm^{-2}K^{-2}$ for 4H-SiC), A is the area of the diode, ϕ_B is the Schottky barrier potential. Hence, a semi-logarithmic plot of the diode current as a function of the bias voltage results in a straight line. As can be seen from Equation (3), the slope of the straight line can be used to calculate the ideality factor and the intercepts give the reverse saturation current. The diode ideality factor and the barrier height thus calculated were 1.4 and 1.3 eV, respectively. It can be noted that the barrier height calculated in the above fashion depends on the spatial homogeneity of the barrier heights. The locations with lower barrier heights will allow more currents compared to the regions with higher barrier heights [19]. The departure from the theoretical diode ideality factor of unity suggests that these Schottky contacts do have spatial non-uniformity of surface barrier height.

The SI 4H-SiC layers (SI-M50) showed a very low capacitance of ≈2 pF which hardly varied with applied bias, confirming its semi-insulating (SI) nature. The n-type epitaxial layers on the other hand showed high sensitivity of depletion capacitance with any change in the applied reverse bias. Figure 5 shows the variation of junction capacitance (C) as a function of the reverse bias voltage for the n-type 4H-SiC epitaxial (n-S50) samples. A straight line fit to the plot of $1/C^2$ as a function of reverse bias voltage gives the built-in potential and the effective doping concentration. The surface barrier height can also be calculated from the C-V measurements using the equations below.

$$\phi_{B(C\text{-}V)} = V_{bi} + V_n \tag{5}$$

where V_n is the potential difference between the Fermi level energy and the bottom of the conduction band in the neutral region of the semiconductor and is given by

$$V_n = k_B T ln \frac{N_C}{N_D} \tag{6}$$

where N_C is the effective density of states in the conduction band of 4H-SiC and is taken equal to 1.6×10^{19} cm^{-3} [15]. The effective doping concentration and the built-in potential was calculated to be 1.1×10^{15} cm^{-3} and 1.4 V, respectively. The $\phi_{B(C\text{-}V)}$ was calculated to be 1.47 eV using Equations (5) and (6) which is slightly higher compared to that calculated from the I-V measurements (1.3 eV). The difference can be attributed to the difference in the underlying mechanism of the two measurements. As has been mentioned earlier, the current flow in the I-V measurements is mostly in the low barrier height regions and hence the barrier height reflects those regions only, whereas the C-V measurements gives the average value of the barrier height for the entire surface area of the Schottky contact. More work is needed to further understand the defects in the material and devise preparation approaches such as surface passivation and edge termination for mitigating these effects [21,22].

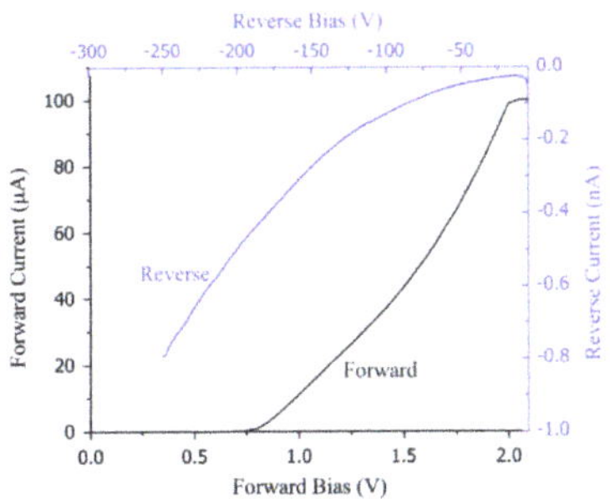

Figure 4. Forward and reverse I-V characteristics obtained for 50 μm n-type 4H-SiC epitaxial layer (n-S50) SBD with 3.8 mm diameter Ni contact.

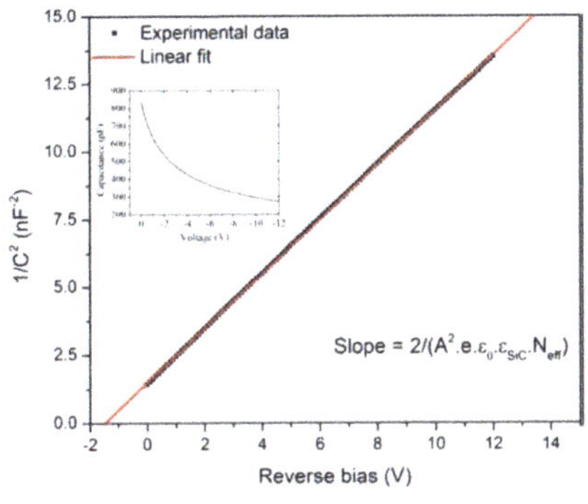

Figure 5. Plot of $1/C^2$ as a function of reverse bias voltage recorded at room temperature for a 50 μm n-type Ni/4H-SiC (n-S50) SBD. The original C-V plot is shown in the inset.

3.2. Quality Evaluation of Epitaxial Layers

The performance of a device obviously depends on the single crystallinity of the epitaxial layer. XRD rocking curve measurement gives very accurate information on the orientation of crystallographic planes. Preferential or defect delineating etching helps in exposing the sites around defects like dislocations, stacking faults, precipitates, and point defects. The width (FWHM) of the XRD rocking curve peak is a measure of the crystalline quality. The lower the FWHM, the higher the crystalline quality [23]. Molten KOH is well-known for its preferential etching on the SiC surface at defect sites [24]. Quality of the semi-insulating 4H-SiC epitaxial layers, used for detector fabrication in this work, was assessed using defect delineating chemical etching in molten KOH followed by XRD rocking curve measurements at the exposed defect sites. For the reflection geometry used in our studies, FWHM of the rocking curve can be calculated [24,25] using the following equation

$$FWHM = \frac{2r_e\lambda^2}{\pi V \sin 2\theta_B} \frac{1}{\sqrt{\gamma}} |C| \sqrt{F_{hkl} F_{\bar{h}\,\bar{k}\,\bar{l}}} \tag{7}$$

where r_e is the classical electron radius, λ is the X-ray wavelength, V is the volume of the unit cell, θ_B is the Bragg angle, $\gamma = \cos(\psi_h)/\cos(\psi_0)$ is the asymmetric ratio, where ψ_h and ψ_0 are the angles between the normal to the crystal surface directed inside the crystal and the reflected and incident directions of X-ray waves, respectively, C is the polarization factor ($C = 1$ for σ polarization and $C = \cos 2\theta_B$ for π polarization), and F_{hkl} is the structure factor with the modulus for (000l) reflection in 4H-SiC (back-reflection geometry) given by equation below [24].

$$|F_{000l}| = 4\sqrt{f_{Si}^2 + f_C^2 + 2f_{Si}f_C\cos(3\pi l/8)} \tag{8}$$

where f_C and f_{Si} are the atomic scattering factors of C and Si atoms, respectively. The scattering factors were calculated using the nine-parameter equation given below [25–27].

$$f = c + \sum_{i=1}^{4} a_i \exp\left[-b_i \sin^2(\theta/\lambda)\right] \tag{9}$$

where a_i, b_i, and c are the atom-specific Cromer–Mann coefficients, which can be found in [26–28]. The FWHM of the (0008) plane reflection was calculated using Equations (7)–(9) and was found to be less than 2.7 arc sec. Figure 6 shows the experimentally obtained rocking curve for (0008) reflection for a semi-insulating 50 μm thick 4H-SiC epitaxial layer [17]. The FWHM of the rocking curve peak was found to be ≈3.6 arc sec, revealing high quality of our epitaxial layer. Structural defect densities were also estimated using a Nomarski microscope. An etch pit density (EPD) of threading screw dislocations (TSDs) was found to be $\approx 1.7 \times 10^3$ cm^{-2}. The concentration of the threading edge dislocations (TEDs) was calculated to be $\approx 1 \times 10^4$ cm^{-2} and basal plane dislocation (BPD) density was found to be ≈70 cm^{-2}.

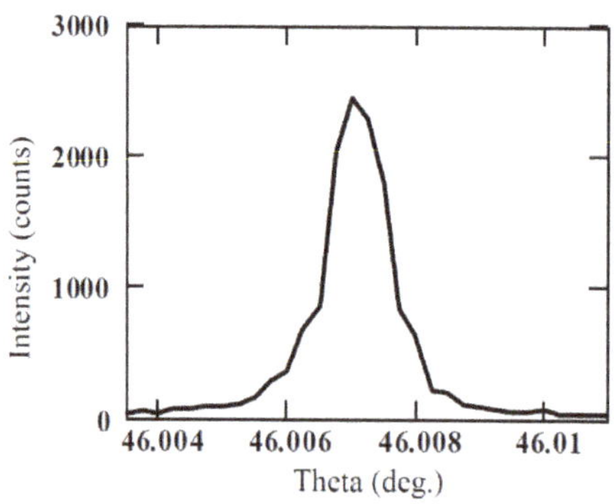

Figure 6. Rocking curve ((0008) reflection) of the 4H-SiC semi-insulating epitaxial layer (SI-M50) used for detector fabrication.

3.3. Radiation Detection Measurements

Most of the following radiation detection measurements were undertaken in the spectrometer configuration shown in Figure 7. The details of the components are described in Section 2.

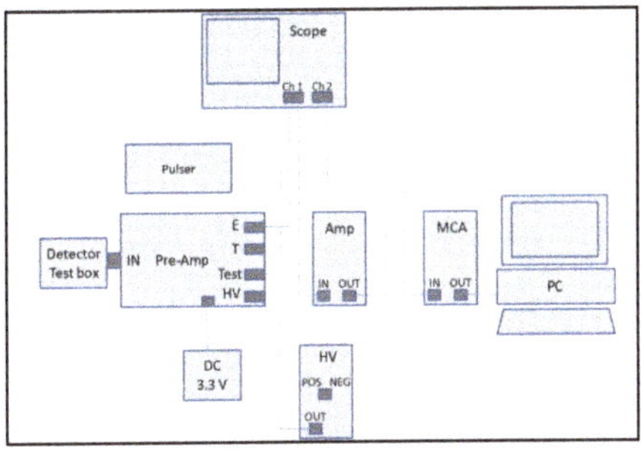

Figure 7. Schematic of the radiation detection set-up.

3.3.1. Electron-Hole Pair Creation Energy Measurements

Electron-hole pair (ehp) creation energy, henceforth designated by ε, determines the energy resolution of a radiation detector as the energy resolution is directly linked to the number of electron-hole pairs created for a single incoming radiation event. The higher the number of ehp, the higher the resolution. A method of iterative determination of the ε value which involves an absolute calibration using a precision pulser to match the alpha peak energy (5486 keV) observed using a high-resolution 4H-SiC n-type epitaxial Schottky detector has been reported in our earlier publication [29]. The alpha particle spectrometer was calibrated electronically by injecting a pulser signal of known pulse height, V_{pulser} (mV), from a precision pulser through a calibrated feed-through capacitor C_{test}, to the preamplifier input. The peak position of the shaped pulses was recorded in a multi-channel analyzer for a set of known pulse-height inputs. The SiC equivalent of the MCA peak positions, E_{pulser} in keV were calculated using the equation given below:

$$E_{pulser} = \frac{V_{pulser} \times \varepsilon \times C_{test}}{q} \tag{10}$$

where q is the electronic charge. A linear regression of the SiC equivalent peak position as a function of MCA channel number was used to calculate the calibration parameters. The detectors used for this study were fabricated on a 20 μm n-type 4H-SiC epilayer. The ε value we obtained using the given procedure was 7.28 eV. The value thus calculated differs from the widely accepted value of 7.7 eV as reported earlier [30]. Rogalla et al. calculated an ε value of 8.4 eV for alpha particles in semi-insulating 4H-SiC [31]. An ε value of 8.6 eV for alpha particles in epitaxial n-type 4H-SiC were reported by Lebedev et al. [32]. Ivanov et al. [33] have reported ε = 7.71 eV for alpha particles in epitaxial n-type 4H-SiC. An ε value of 7.8 eV has been reported by Bertuccio and Casiraghi for 59.5 keV gamma rays [34].

3.3.2. Minority Carrier Diffusion Length Measurements

Minority carrier diffusion length is the average distance a minority carrier traverses before it recombines. Higher minority carrier diffusion length obviously enhances the detection properties by reducing the effect of ballistic deficit [35]. The minority carrier diffusion length can be indirectly calculated by observing the variation of charge collection efficiency of detectors for ionizing radiations like alpha particles with reverse bias voltage. Charge collection efficiency (CCE) is defined as the ratio of charge collected by the collecting electrode at a particular bias to the maximum collected charge, assuming all the charge carriers have been received by the collecting electrode. The collected charge is generally calculated by integrating the current signal received at the input of a charge-sensitive pre-amplifier. Alternatively, the CCE can also be calculated from alpha spectroscopic measurements. The MCA peak position E_p due to a monoenergetic alpha source can be predicted in a properly calibrated alpha spectrometer, assuming that all the charge carriers created by that particular energy have been received by the collecting electrode. The CCE can then be determined for any MCA peak at E_α by calculating the ratio E_α/E_p. Figure 8 shows the variation of observed charge collection efficiency (CCE_{obs}) of a n-type Ni/4H-SiC SBD as a function of reverse operating bias voltage [36,37]. By noting that the movement of the charge carriers in rectifying junctions can be due to diffusion as well as drifting, the contribution of each mechanism to CCE can be calculated using a proper model. The contribution of the drift $CCE_{depletion}$ and diffusion $CCE_{diffusion}$ related charge collection to the CCE_{obs} was calculated using a drift-diffusion model [38] summarized below in Equation (11).

$$CCE_{theory} = \frac{1}{E_p}\int_0^d \left(\frac{dE}{dx}\right)dx + \frac{1}{E_p}\int_d^{x_r}\left[\left(\frac{dE}{dx}\right)\times \exp\left\{-\frac{(x-d)}{L_d}\right\}\right]dx \qquad (11)$$
$$= CCE_{depletion} + CCE_{diffusion}$$

where d is the depletion width at the particular bias, $\frac{dE}{dx}$ is the electronic stopping power of the alpha particles calculated using SRIM [39], x_r is the projected range of the alpha particles with energy E_p. We fitted the CCE_{theory} values to the CCE_{obs} values considering L_d, the minority carrier diffusion length, as a free parameter. The L_d value corresponding to the best fit was returned as the average minority carrier diffusion length. For the present SBD, the average value of L_d was found to be ≈18.6 µm. From Figure 8 it was also observed that the $CCE_{diffusion}$ values dominate considerably over that of $CCE_{depletion}$ up to a reverse bias of −30 V. At even higher bias voltages the depletion width becomes equal or more than the projected range of alpha particles (≈18 µm in SiC for 5486 keV alpha particle) and hence charge collection was mainly due to the drifting of charge carriers within the depletion width. Hence, above bias voltage of −70 V, the $CCE_{depletion}$ matched the CCE_{obs} values.

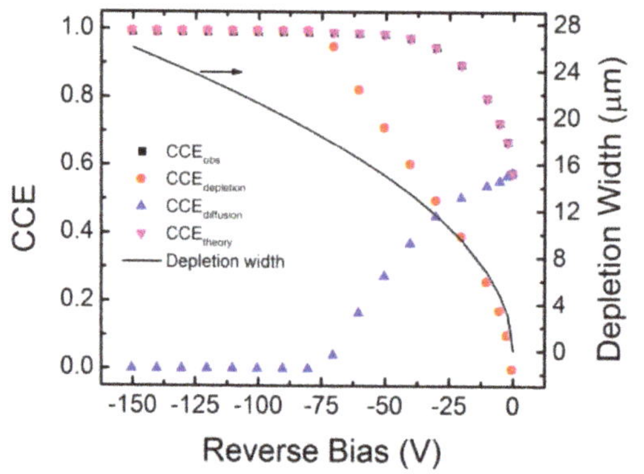

Figure 8. Variation of CCE_{obs} and CCE_{theory} with reverse bias voltage for an n-type Ni/4H-SiC (n-S20) SBD. The contributions of $CCE_{depletion}$ and $CCE_{diffusion}$ to the total CCE_{obs} are also plotted. The solid line shows the variation of depletion width with bias.

3.3.3. Alpha Particle Pulse-Height Spectroscopy (PHS)

The n-type 4H-SiC epitaxial layer detectors were put to test as an alpha particle detector for precise energy measurements. A ^{241}Am alpha particle source was used as the test source. Figure 9 shows a typical pulse-height spectrum obtained from a n-type 4H-SiC SBD [40]. As is evident from the Figure 9, the detector clearly resolved the three primary alpha particles emitted by a ^{241}Am alpha source. The percentage energy resolution (in terms of full width at half maxima or FWHM) for the 5486 keV line was calculated to be 0.29% with a ≈100% charge collection efficiency. The observed alpha energy resolution is so far the best-reported value to date in the literature. At this point, it is also worth discussing the evolution of 4H-SiC epitaxial layers as alpha particle detectors. The idea to use 4H-SiC was pioneered by Babcock and Chang [1]. Ruddy et al. [41] reported a percentage energy resolution of 5.8% and 6.6% for a deposited energy of 294 and 260 keV alpha particles, respectively.

It can be noted that the authors used a collimated ^{238}Pu source and circular diode contacts of 200 and 400 μm. Later on, Ruddy et al. also reported [42] an energy resolution of 5.7% for a deposited energy of 89.5 keV alpha particles from a 100 μm collimated ^{148}Gd source in similar detectors with comparatively larger Schottky contact diameter of 2.5, 3.5, 4.5, 6.0, and 10 mm thick 4H-SiC epitaxial layers. In another work [8], Ruddy et al. reported an energy resolution close to 46 keV (≈0.8%) for alpha particles from a ^{238}Pu source and 41.5 keV (≈1.3%) for alpha particles from a ^{148}Gd source for devices with an aluminum guard ring. Ivanov et al. [43] reported an energy resolution of 20 keV (≈0.4%) in the energy range 5.4–5.5 MeV. In yet another work, Ruddy et al. [44] reported an energy resolution of 20.6 keV (≈0.4%) for ^{238}Pu alpha particles. In our earlier work [20] we reported an energy resolution of 2.7% for 5486 keV alpha particles in 50 μm thick n-type Ni/4H-SiC detectors. The energy resolution mentioned in the above works were found to be primarily dependent on the defect type and concentrations within the 4H-SiC epilayers. The nature of defects, which controls the detector properties, will be described in detail in a later section.

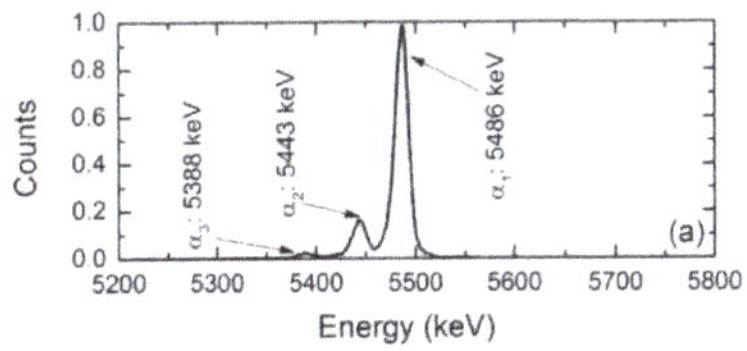

Figure 9. A pulse-height spectrum obtained for a 20 μm n-type 4H-SiC (n-S20) SBD and ^{241}Am alpha source.

3.3.4. Low Energy Gamma Spectroscopy

Absolute measurement of photo-responsivity and probing of physical construction of photonic sensors can be very effectively done using synchrotron light sources. N-type 4H-SiC epitaxial layer detectors fabricated at UofSC were studied at NSLS at BNL for detection of low energy X-rays. The results were compared to a commercial off-the-shelf (COTS) SiC UV photodiode by IFW optronics GmbH (Jena, Germany), model JEC4 which was known to be the best commercially available for such applications [45]. An X-ray spectrometer for such a low energy spectral range is not available commercially. Figure 10 shows the responsivity of one of our detectors and a IFW JEC4 SiC UV photodiode to soft X-ray energy ranges biased at 250 and 120 V, respectively [17]. The following results were derived using a statistical analysis of these data based on energy-dependent X-ray attenuation lengths [46].

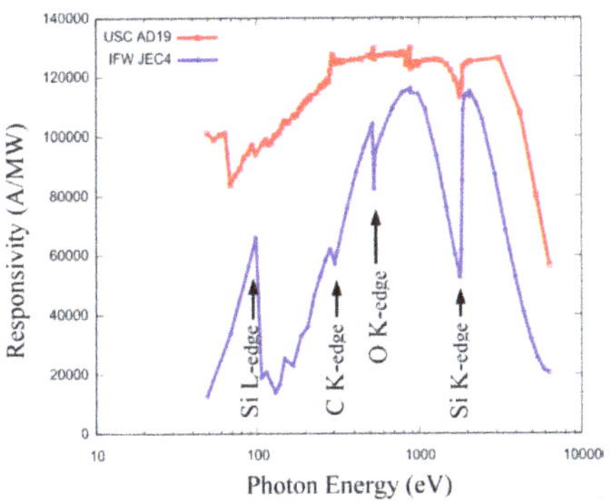

Figure 10. Measured responsivity of a 4H-SiC n-type epitaxial device biased to 250 V and an IFW JEC4 photodiode biased to 120 V.

Responsivity measurements were carried out using the U3C [45] and X8A [46] lines by recording successive measures of photocurrent in response to a high flux, mono-energetic beam of photons in a well-calibrated silicon sensor (with known responsivity) and in the sensor of interest (Figure 11a). Dead layers and a limited active volume thickness led to responsivity that varies greatly with photon energy. Further, edges were also apparent in the responsivity curve, arising from discrete atomic transitions. Edges associated with silicon and carbon is clearly observed in the data, providing a quantitative measure of the composition and dimension of the active and dead layers [46]. The general feature of a steep decrease starting at 2–3 keV provides information on active layer thicknesses, which is deduced to be 21 μm in our detector compared to roughly 6 μm for the JEC4 diode [47]. Due to the higher active layer thickness our sensor chip showed significant improvement of responsivity in the few keV range compared to COTS SiC UV photodiode. Our detector has shown much higher response in the low energy part of the spectra as well, which could be attributed to a much thinner dead/blocking layer, deduced from the responsivity curve to result solely from the 10 nm thick nickel layer (which leads to the pronounced edge at 70 eV). In comparison, the JEC4 diode has been found to include a significant oxidation and inactive SiC layer on the order of 100 nm each, which limits responsivity at low photon energies [47]. It should be noted that the JEC4 diode is intended for UV detection, for which it is well suited. The significant dead layers are likely due to passivation, which may be preferred over reducing the thickness of dead layers on the active face of the sensor.

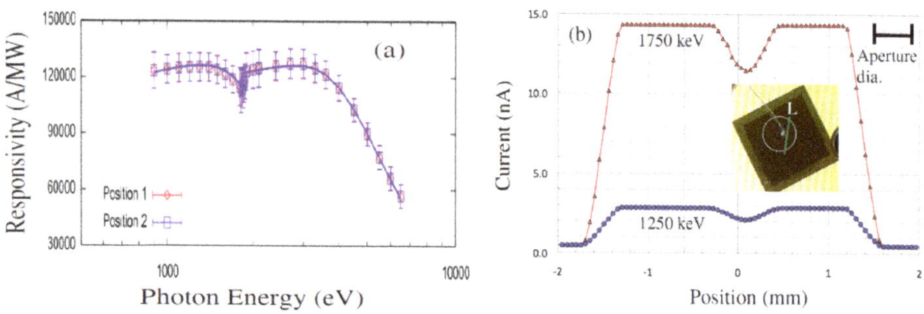

(a) (b)

Figure 11. (**a**) Responsivity of the detectors on 4H-SiC n-type epitaxial layer at two different locations and (**b**) surface scan profiles along line L obtained to assess the detector's uniformity.

Our detectors also exhibited very good spatial uniformity in measured responsivity. Figure 11b shows responsivity at two different locations and line scan profiles for two different X-ray energies. Note that the decrease of the detector's current at about 0 mm (Figure 11b) is due to the crossing of the location of wire bonding and not due to the detector's imperfection. The SiC detectors were connected to low noise front end electronics developed in-house for pulse-height spectroscopy (PHS). Pulse height spectra measurements showed high resolution of our 4H-SiC detectors in detecting 59.6 keV gamma rays from ^{241}Am. Figure 12 shows a pulse height spectrum recorded using an ^{241}Am radiation source with the detector biased at 250 V, with an FWHM of 1.2 keV (2.1%) at 59.5 keV, which is comparable to the resolution achieved using high quality CdZnTe detectors [48,49].

3.4. Electronic Noise Measurements

The energy resolution of nuclear energy spectrometers is dependent on the noise of the detector and associated electronic modules in the spectrometer and the pre-amplifier in particular. Noise is defined as any statistical fluctuation in currents measured in the detector or associated electronics which constitute a signal. The most appropriate way to monitor the noise is to capture the pulses from a standard pulser along with the pulses produced by a detector due to the incoming radiation. The pulse-height spectrum then reveals a peak due to the pulser with the actual radiation peaks.

The width (FWHM) of the pulser peak then gives the idea of the overall noise of the spectrometer ($FWHM_{total}$).

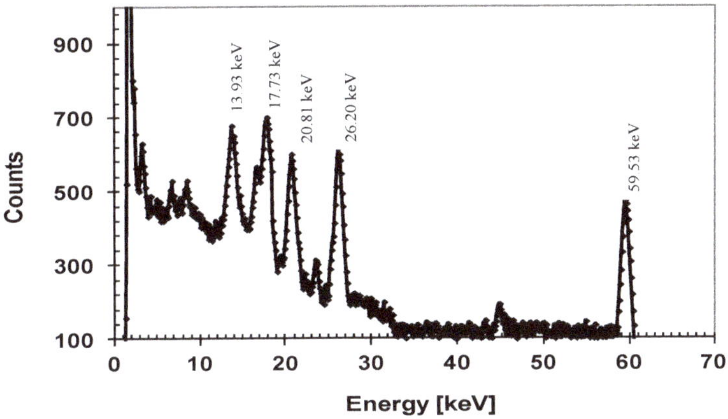

Figure 12. ^{241}Am spectrum of a 4H-SiC detector (8.0 mm^2) at 300 K, 250 V bias, and 4 µs shaping time.

Figure 13 shows the results of a typical noise monitoring measurement [36,37]. The energy resolution of the detector could be seen to improve with increase in bias voltage up to 100 V reverse bias because of the increase in depletion width (active volume of the detector) and lowering in capacitance. The energy resolution beyond 100 V was seen to deteriorate with increase in bias. The increase in leakage current was ruled out as an explanation as it could be seen from the figure that the pulser noise did not increase at all. A possible reason behind the deterioration of the resolution is incorporation of the threading dislocation as the depletion width approaches towards the epilayer–substrate interface with the increase in reverse bias. The epilayer–substrate interfacial region is prone to have a larger threading type dislocation concentration, which propagates from the substrate to the epilayer [36].

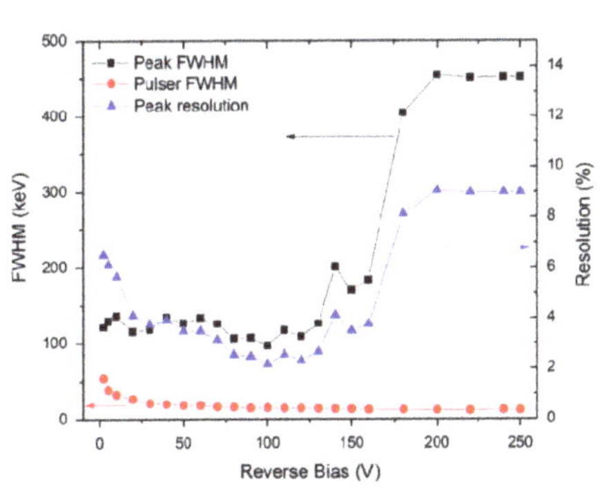

Figure 13. Variation of 5.48 MeV alpha peak FWHM, pulser peak FWHM, and alpha peak percentage resolution as a function of detector bias voltage.

For a superior quality n-S20 detector, the $FWHM_{total}$ was found to be 19.8 keV for 5486 keV alpha particles. The contribution from the noise from the front-end electronics ($FWHM_{elec}$), and the detector leakage current ($FWHM_{leakage}$), can be found from the width of the MCA pulser peak recorded with the detector plugged in and biased. The collective broadening due to $FWHM_{elec}$ and $FWHM_{leakage}$ for this detector was found to be 15.9 keV. The other contributions to the $FWHM_{total}$ are from the statistical fluctuation in the number of ehps produced by ionizing radiation $FWHM_{stat}$, and broadening due to variation of energy due to the entrance window, the angle of incidence, self-absorption in the source, etc. $(FWHM_{other})$. All these factors along with the intrinsic detector resolution $FWHM_{\mathrm{det}}$ are related to the ultimate peak broadening through the relation given in Equation (12). $FWHM_{stat}$ and $FWHM_{other}$ were calculated in this detector and found to be 5.3 and 0.44 keV, respectively [29]. The intrinsic detector resolution was calculated from Equation (12) and found to be 10.5 keV.

$$FWHM_{total}{}^2 = FWHM_{\mathrm{det}}{}^2 + FWHM_{leakage}{}^2 + FWHM_{stat}{}^2 + FWHM_{elec}{}^2 + FWHM_{other}{}^2 \quad (12)$$

The electronic noise has various sources such as detector leakage current and capacitance, and input FET (field effect transistor in the pre-amplifier) noise. The contribution of the different sources to the signal-to-noise ratio is dependent on the filtering or shaping operation and in particular the shaping time (except the FET noise). A clear understanding of the electronic noise is thus very essential to proper tuning of the spectrometer and a pioneering work in this area has been conducted by Bertuccio and Pullia [50]. The electronic noise was expressed in terms of equivalent noise charge (ENC) and plotted as a function of the shaping time τ of the shaping amplifier. The plots were then fitted to Equation (13) below using a least square estimation method.

$$ENC^2 = \left(aC_{tot}^2 A_1\right)\frac{1}{\tau} + \left[\left(2\pi a_f C_{tot}^2 + \frac{b_f}{2\pi}\right)A_2\right] + (bA_3)\tau \quad (13)$$

where C_{tot} is the total input capacitance, A_1, A_2, and A_3 are constants depending on the shaping network response, a is the coefficient of white series noise contribution due to the thermal noise of the FET channel, a_f is the coefficient of the FET $1/f$ noise, b_f is the dielectric noise coefficient, and b is the coefficient related to the sum of the white parallel noise due to the shot noise and the detector leakage current.

Figure 14a,b shows the variation of ENC with shaping time without and with the detector (20 µm n-type Ni/4H-SiC (n-S20) SBD) connected [37]. The contributions from the three different terms were plotted separately. The minimum noise without the detector was observed to correspond to a shaping time value between 1 and 2 µs. The same shifted to a higher range of shaping time (between 3 and 6 µs) when the detector was plugged in. It can also be seen that the white parallel noise increased almost by a factor of five after connecting the detector and the pink noise marginally increased for any given τ after connecting the detector. In contrast, the white series noise increased by an order of magnitude when the detector was connected. The increase in white parallel noise can be attributed to the increase in the leakage current (from the detector) and the increase in white series and parallel noise is supposedly due to the increase in input capacitance when the detector is plugged in.

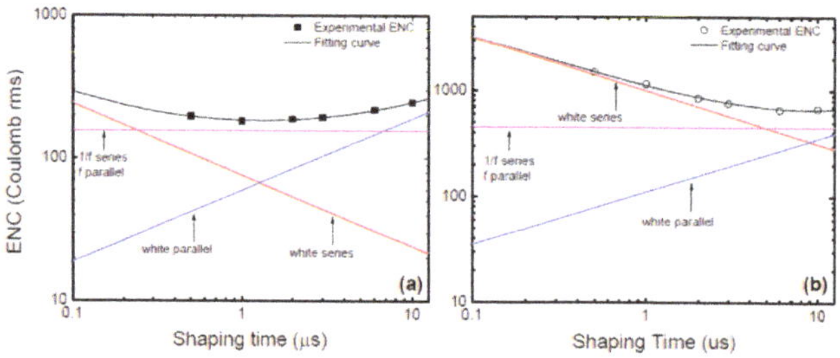

Figure 14. Variation of equivalent noise charge as a function of shaping time constant (**a**) without the detector connected, (**b**) with the detector connected.

3.5. Defect Level Characterization

3.5.1. Morphological Defect Study using Electron Beam Induced Current (EBIC) Studies

The SI (SI-M50) and n-type 4H-SiC (n-M50) epitaxial layer samples have been studied for the presence of morphological defects [51]. Figure 15 shows the morphological defects that were observed in the n-type epitaxial layer samples followed by KOH etching, optical microscopy, and EBIC studies [52]. While the n-type epitaxial layers showed features like comet tails, pits, hillocks, triangular defects, and step bunching, the SI epitaxial layers showed the presence of carrot defects only. The hillocks were proposed to originate from foreign impurities and silicon precipitates. Silicon can also have pits if it evaporates during the epitaxial growth. Triangular defects indicate the inclusion of 3C-SiC and Shockley-type stacking faults nucleating on micropipes and elementary threading screw and edge dislocation [53]. The presence of these defects is believed to increase the leakage current upon application of high electrical fields to the devices [54]. Pits and step bunching types of morphological defects are not believed to influence the leakage current but can interfere with proper functioning of the Schottky contacts in case the epilayer surface is not smooth enough due to their presence [54]. Figure 16 shows an EBIC image of an n-type 4H-SiC epitaxial layer. The typical signatures of threading dislocation type defects could be seen as black spots [55,56]. The EBIC features were mapped on to the morphological defect images and it was correlated that the dark spots are the signatures of the comet tail morphological defects in the n-type epitaxial layers. The superior samples did not show any presence of morphological defects.

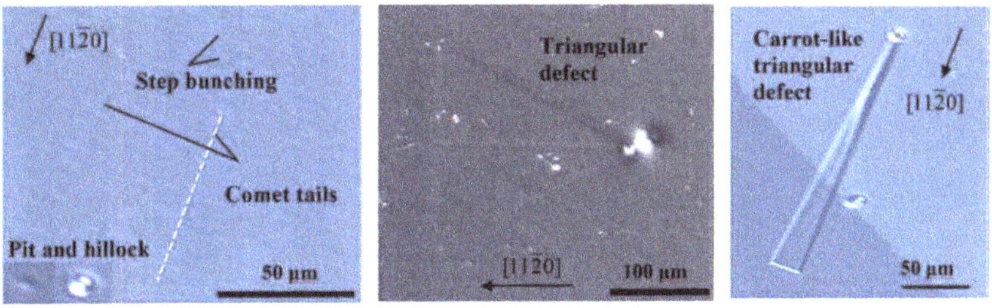

Figure 15. Microscopic image of the morphological defects revealed after KOH etching in n-type 4H-SiC samples.

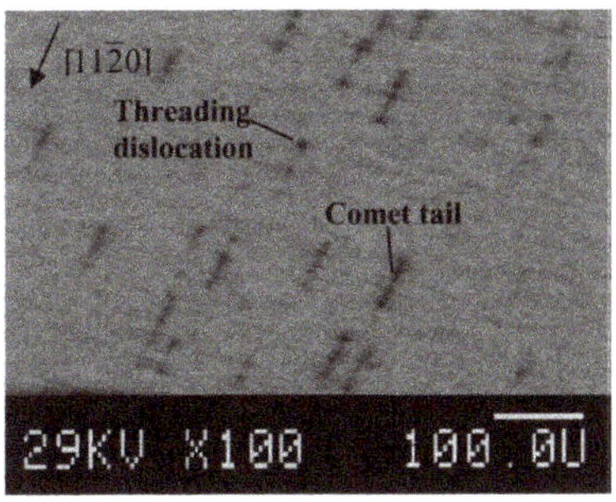

Figure 16. Electron beam induced current spectroscopy (EBIC) image of a n-type 4H-SiC epitaxial layer.

3.5.2. Thermally Stimulated Current (TSC) Measurements

Thermally stimulated current measurement is yet another sensitive technique to study defects in semi-insulating (SI) as well as conducting samples. Figure 17 shows TSC spectra obtained from an n-type 4H-SiC epitaxial layer reverse biased at two different voltages, 4 and 12 V [56]. The spectra were acquired with a heating rate of 15 K/min. Four discernable peaks were observed in both the TSC spectra and were numbered 1 to 4. Peak#1 has the most prominent presence with maximum temperature T_m = 109 K and an activation energy ≈0.25 eV as estimated from the Arrhenius plot (not shown). This peak can be associated with shallow acceptor-like levels situated at 0.25 eV from the valence band edge and may be related to Al- and/or B-impurities as well as to their complexes with intrinsic defects [56]. It can be noticed that the intensity of the peak#1 showed voltage dependence, which implies that the deep level centers are distributed over a thickness of 4 μm in the epilayer as 4 μm is the depletion width achieved at 12 V. The concentration of the associated defects was estimated to be ≈7 × 10^{13} cm^{-3}, assuming uniform distribution of deep level centers in the depletion region. Equation (14) was used to calculate the trap concentrations, N_T:

$$N_T = \frac{Q}{A}\sqrt{\frac{2N_d}{q\varepsilon_{SiC}\varepsilon_0(V_{bi} + V_a)}} \tag{14}$$

where, Q is the total charge emitted by the given trap which in turn is determined by the area under the corresponding peak, A is the contact area, N_d is the effective doping concentration, V_{bi} is the built-in potential, V_a is the applied bias voltage, q is the electronic charge, ε_{SiC} is the dielectric permittivity of SiC, and ε_0 is the dielectric permittivity of vacuum.

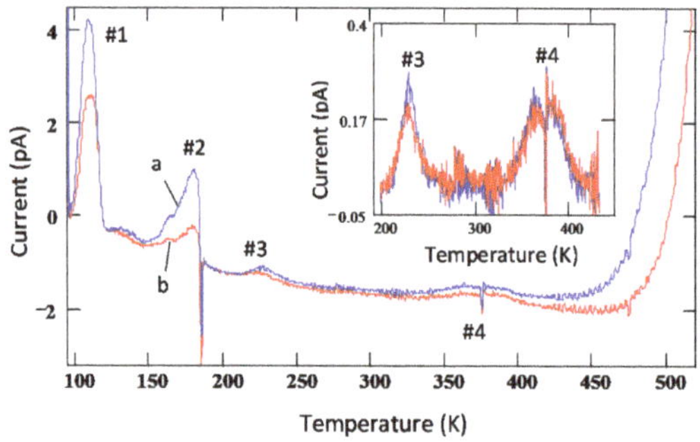

Figure 17. Thermally stimulated current (TSC) spectra obtained using an n-type 4H-SiC (n-M50) epitaxial layer at two different reverse bias voltages: (**a**) 12 V and (**b**) 4 V.

It can be noted here that being a shallow defect level, the trap center corresponding to peak #1 is not expected to cause significant trapping/polarization even though its concentration is relatively high. The intensity of peak#2 was also observed to increase with reverse bias voltage. However, it depended on other conditions such as pumping time. Overnight pumping, since the exposure of the TSC chamber and the sample to air, resulted in decrease of the intensity of peak#2 by a factor of two to three. This suggests that peak#2 could have contributions from levels/dipoles produced by adsorption of residues in the vacuum chamber (such as water) onto the surface of the sample. Additionally, peak#2 was always distorted by the negative spike, the origin of which could not be explained. The intensities of peaks #3 and #4 do not show voltage dependence of the peak intensities as can be observed when zoomed-in (inset in Figure 17). This is an indicative of the fact that the traps associated with peaks #3 and #4 are located mostly near the metal-semiconductor interface and not in the interior of the epitaxial layer. The activation energies for peaks #3 and #4 could not be determined as the TSC signals were too weak. However, the traps corresponding to peaks #3 and #4 could be identified using their maximum temperatures and previously reported data in similar samples [12,14,56]. The peak#3 (T_m ≈226 K) can be assigned to D-center, a boron (B) related defect, boron at C-site (B_C) or boron at Si site (B_{Si}), and carbon vacancy V_C [57], whereas peak#4 can be assigned to intrinsic defects such as IL2 center [58,59].

Figure 18 shows TSC spectra obtained for a semi-insulating 4H-SiC epitaxial layer (SI-M50) sample at 0 V applied bias (higher bias resulted in large leakage currents for these samples) [14]. The thermally stimulated current at 0 V is due to a thermoelectric effect caused by small temperature gradient between the front and the back surface of the samples [60]. Five peaks of either polarity were observed and named as A–E. The positive peaks represent hole traps and the negative peaks represent electron traps as the polarity of the thermoelectric effect current is sensitive to the type of carrier. The Arrhenius plots of the designated traps, except for E, are shown in Figure 19. The trap parameters, calculated from the Arrhenius plots, and their identity are tabulated in Table 2. Due to high stray currents at high temperatures, peak E could not be used for Arrhenius plots.

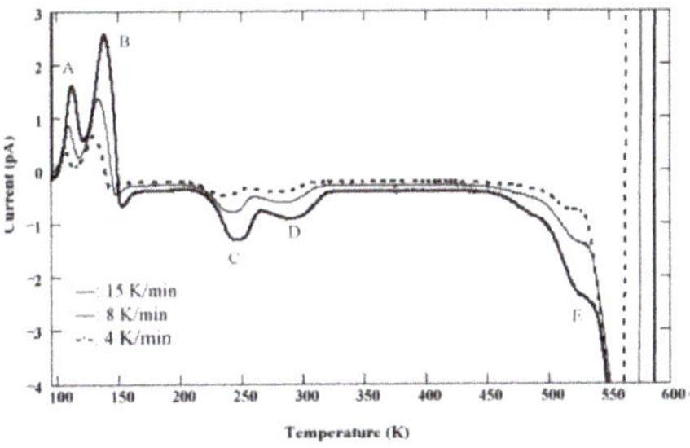

Figure 18. TSC spectra obtained using a SI 4H-SiC (SI-M50) epitaxial layer for three different heating rates.

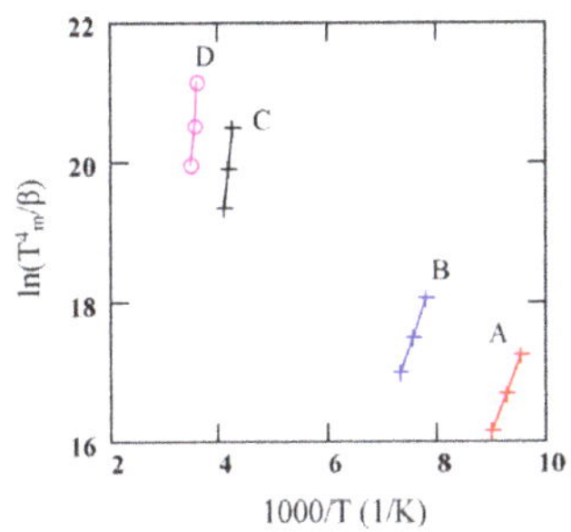

Figure 19. Arrhenius plots obtained from TSC spectra of a SI 4H-SiC (SI-M50) epitaxial layer at 0V bias. voltage.

Table 2. Defect parameters acquired from TSC spectroscopy for a SI Ni/4H-SiC sample.

Peak #	T_m (K)	Activation Energy (eV)	Capture Cross-Section (cm^2)	Possible Trap Identity
A	105–112	0.19	$\approx 10^{-18}$	Al, B, L-center [59,61]
B	128–137	≈0.2	-	
C	236–247	0.57	$\approx 2 \times 10^{-16}$	$Z_{1/2}$ [59,61,62]
D	280–290	0.82–0.87	$\approx 10^{-13}$	IL_1 [58]
E	≈525	-	-	Intrinsic defects

3.5.3. Deep Level Transient Spectroscopy (DLTS) Measurements

Deep level capacitance transient spectroscopy is a very sensitive technique to study deep level defects in semiconducting Schottky or p-n junction devices. Figure 20 shows a DLTS spectra obtained for a 50 μm thick n-type Ni/4H-SiC (n-S50) epitaxial Schottky barrier detector in the temperature range 85–790 K [63]. Six well-resolved peaks were observed. All the peaks, except the deepest one (peak#6), were identified using the existing literature. The Arrhenius plots are shown in Figure 21 and the extracted trap parameters are tabulated in Table 3.

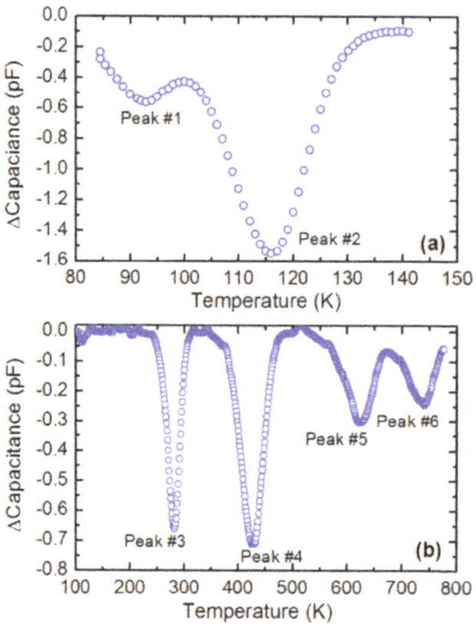

Figure 20. A typical deep level transient spectroscopy (DLTS) spectra obtained using the 50 μm n-type Ni/4H-SiC (n-S50) epitaxial Schottky barrier detector in a temperature range of (**a**) 80–140 K and (**b**) 80–800 K.

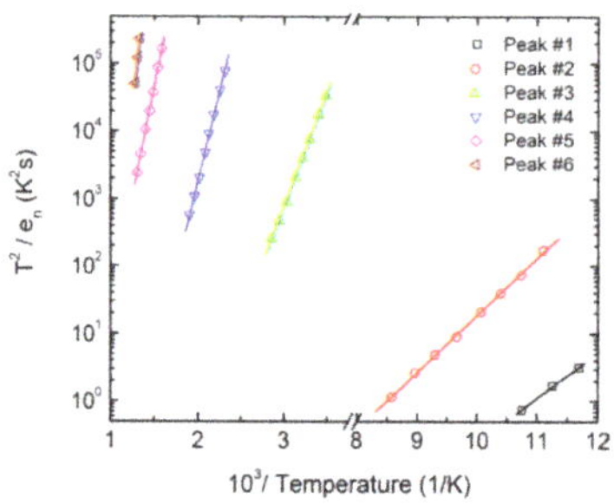

Figure 21. Arrhenius plots obtained for peaks #1–#6 corresponding to the DLTS spectra shown in Figure 20.

Table 3. Defect parameters acquired from DLTS spectroscopy for a 50 μm n-type Ni/4H-SiC (n-S50) epitaxial Schottky barrier detector.

Peak #	Capture Cross-Section (cm^2)	Activation Energy (eV)	Trap Concentration (cm^{-3})	Possible Trap Identity
1	4.13×10^{-15}	$E_c - 0.13$	1.3×10^{13}	Ti(h) [62,64–67]
2	2.50×10^{-15}	$E_c - 0.17$	3.6×10^{13}	Ti(c) [62,64–67]
3	3.36×10^{-15}	$E_c - 0.67$	1.7×10^{13}	$Z_{1/2}$ [66–69]
4	3.73×10^{-15}	$E_c - 1.04$	2.1×10^{13}	EH_5 [70–73]
5	3.22×10^{-17}	$E_c - 1.30$	7.9×10^{12}	Ci1 [71]
6	1.53×10^{-11}	$E_c - 2.40$	5.6×10^{12}	Unidentified

Figure 22 shows DLTS spectra obtained from a 20 μm Ni/4H-SiC epitaxial (n-S20) SBD using a −2 to 0 V pulse with a pulse width of 1 ms on the temperature range 84 to 750 K. Figure 22a was recorded using correlator delays of 100, 50, 20, and 10 ms and Figure 22b was recorded using 0.2, 0.1, 0.05, and 0.02 ms for a shorter temperature range. The initial spectra showed four peaks; however, after gaussian deconvolution a fifth peak labeled peak #4 was observed between peaks #3 and #5. The most dominant peak is clearly peak #2.

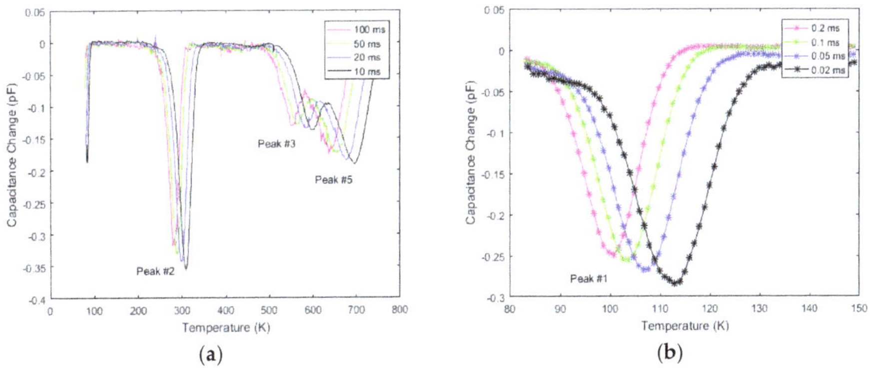

Figure 22. DLTS spectra of the detector obtained using the highest (**a**) and lowest (**b**) sets of initial delays over the temperature range of 84–750 K.

From the Arrhenius plot (not shown), the activation energy and capture cross-section of each peak were obtained. The associated defect parameters are tabulated in Table 4.

Table 4. Defect parameters acquired from the DLTS spectra for a 20 μm n-type Ni/4H-SiC epitaxial (n-S20) Schottky barrier detector.

Peak #	Capture Cross-Section (cm^2)	Activation Energy (eV)	Trap Concentration (cm^{-3})	Possible Trap Identity
Peak #1	1.742×10^{-14}	$E_c - 0.170$	2.05×10^{12}	Ti(c)
Peak #2	3.970×10^{-15}	$E_c - 0.647$	2.67×10^{12}	$Z_{1/2}$
Peak #3	1.721×10^{-14}	$E_c - 1.385$	4.89×10^{11}	Ci1
Peak #4	1.055×10^{-15}	$E_c - 1.316$	6.39×10^{11}	EH_6
Peak #5	2.672×10^{-15}	$E_c - 1.537$	1.21×10^{12}	EH_7

Peak #1 is well established to correlate with the transition of substitutional titanium (Ti) in the cubic lattice site, Ti(c), from the +3 charge state to +2 [64] and is well known to appear in SiC as a side effect of the growth process [61,64,66]. The upward bending of the left side tail of the peak for the lowest two correlators suggest the presence of the Ti(h) defect at approximately $E_c - 0.12$ eV; however, its peak was not observed in the temperature range used. Peak #2 is the $Z_{1/2}$ center which appears in all 4H-SiC samples and has been strongly correlated to carbon vacancies as established by several annealing studies using DLTS and EPR [74–76]. Theoretical calculations and EPR measurements suggest that the identity of the three levels could be the (−2/0) transition of the cubic and (−1/0) transition of both the cubic and hexagonal site carbon vacancies [72,73,76,77]. Peak #3 is identified as Ci1 which is suspected to originate from chlorine impurities introduced during the growth process to compensate for silicon droplet formation [61,71]. Peaks #4 and #5 were labeled as EH_6 and EH_7 and are commonly grouped together as the single peak $EH_{6/7}$ due to their close proximity [14,51,52,56,66,75,78–81]. Its concentration has a one to one correlation with $Z_{1/2}$ suggesting its relation to the carbon vacancy. Further EPR measurements suggest that EH_6 is the +1 donor state and EH_7 is the +2 state [77] and this is supported by theoretical calculations as well [72,73,79]. In contrast to the n-S50 samples, the n-S20 sample did not show the presence of Ti(h), EH_5, and the unidentified defect situated at 2.4 eV below the conduction band edge. However, n-S20 samples did show the presence of EH_6 and EH_7 defect levels which were not encountered in the n-S50 samples.

3.5.4. Isochronal Annealing Studies

To study the defect dynamics, i.e., transformation or disappearance of defects because of atomic motion under the influence of temperature, isochronal annealing experiments were carried out on 50 μm n-type Ni/4H-SiC (n-S50) SBDs [82]. As mentioned earlier in Section 2, the samples were annealed at a particular temperature for 30 min followed by DLTS measurements. The annealing treatments were carried out in the temperature range 100–800 °C. Figure 23 summarizes the DLTS results obtained after each annealing stage. For the sake of simplicity, the defect levels with activation energy above room temperature are shown in the figure. As is evident from the DLTS spectra, all the deep level defects were very much stable up to an annealing temperature of 800 °C. The lesser apparent defects as have been observed previously [82] are also summarized as follows. The capture cross-sections of the trap centers Ti(c), $Z_{1/2}$, and EH_5 were reduced by an order of magnitude when the samples were annealed at a temperature of 400 °C. The respective defect densities were observed to follow a similar trend throughout the isochronal annealing studies.

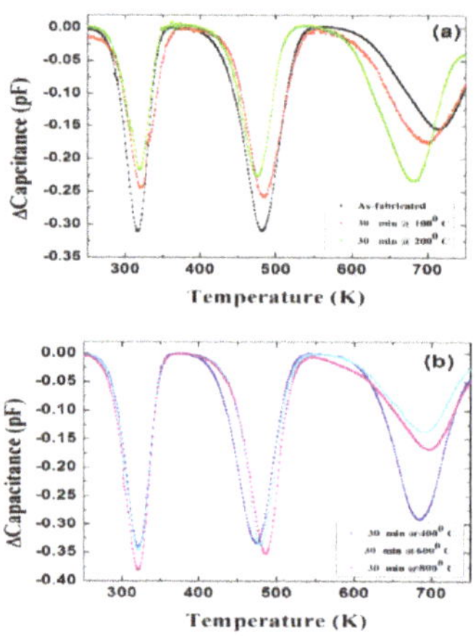

Figure 23. DLTS spectra obtained in a temperature range 250–750 K: (**a**) as-fabricated and annealed at 100 and 200 °C; and (**b**) annealed at 400, 600, and 800 °C.

3.5.5. Correlation of Detector Performance with Deep Level Defects

Deep level capacitance transient spectroscopy was carried out in three n-S20 type samples. The detector performance was also studied for alpha radiation. Figure 24 shows DLTS spectra in the capacitance mode for these three samples named as AS1, AS2, and AS3. Figure 25 shows the alpha pulse-height spectra for the same detectors. The energy resolution of the detectors could be seen to vary although they were fabricated in one batch under a similar condition. The detector AS1 showed the best resolution (0.29%), followed by AS2 (0.38%) and AS3 (0.96%) for 5486 keV alpha particles. From the DLTS spectra it is evident that the best detector AS1 almost did not show Peak#1 which is associated with the $Z_{1/2}$ defect. The trap concentration related to the other peaks were also orders of magnitude lower (except for Peak#3 which could not be properly identified) when compared to that in AS2 and AS3 [40]. Between AS2 and AS3, AS2 exhibited better detector performance which is also corroborated by the fact that the capture cross-sections of the $Z_{1/2}$, Ci1(Peak#3), and $EH_{6/7}$ (Peak#4) defects were at least one order of magnitude lower compared to the detector AS3. The presence of the Ci1 defect did not seem to interfere with the detector performance as the detector AS2 has much larger concentration and cross-section of the Ci1 defect compared to the other two samples. On the other hand, $Z_{1/2}$, which was identified as carbon vacancies and situated at 0.6 eV below the conduction band edge, evidently deteriorates the performance of detectors.

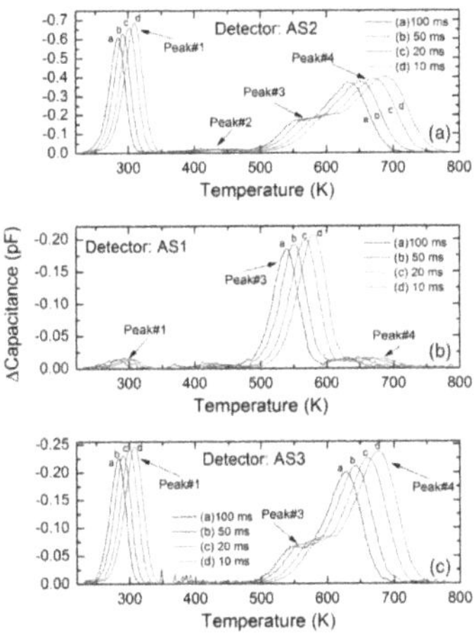

Figure 24. Capacitance-DLTS spectra obtained in the temperature range 250–750 K for (**a**) AS2, (**b**) AS1, and (**c**) AS3. The detectors were fabricated on n-S20 type epitaxial layers.

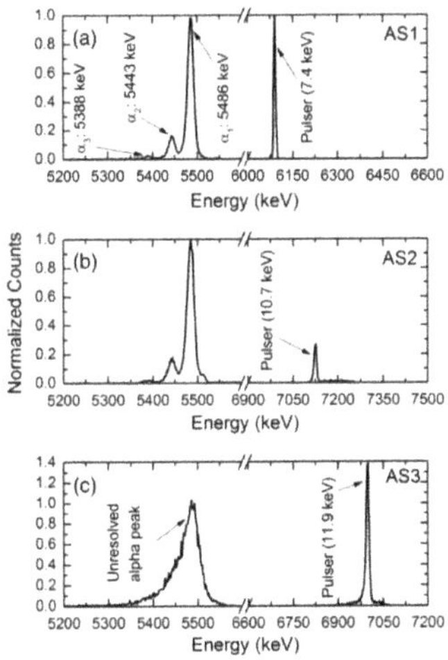

Figure 25. Pulse height spectra obtained using a ^{241}Am point alpha source for detectors (**a**) AS1, (**b**) AS2, and (**c**) AS3.

Hence, although the detectors were fabricated from superior quality wafers taken from the same parent wafer and batch processed under exactly similar conditions, it can be seen that their performance as radiation detectors varied significantly. The relative difference in performance was seen to be related with the presence of defects suggesting inhomogeneous distribution of such defects in the parent wafer.

4. Discussion

I-V measurements on the SI epitaxial layers showed the formation of contacts with asymmetric behavior with respect to the polarity of the applied bias. These samples also showed the influence of trapping centers on the I-V patterns in the forms of steps. The n-type samples on the other hand formed very effective Schottky contacts and did not show any influence of trapping centers on the I-V curves obtained at room temperature. The typical barrier height of such Schottky contacts were found to be of the order of 1.3 eV with diode ideality factors mostly greater than 1 indicating spatial non-uniformity of barrier height. Temperature dependent I-V characteristic measurements also revealed that the effective area through which actual current flow takes place on the Ni/4H-SiC interface is at least an order of magnitude less than the actual geometric contact area.

The semi-insulating samples showed a very little capacitance (≤2 pF) and hardly showed any variation with bias voltages. The n-type epitaxial layer samples as a Schottky diode, showed a comparatively high capacitance value of typically 800 pF at 0 V reverse bias. The effective doping concentration was calculated to be of the order of 1×10^{15} cm^{-3}, from the C-V measurements. The typical built-in potential was calculated to be 1.4 V. Surface barrier heights were also calculated from the C-V measurements and were found to be in the order of 1.47 eV which is slightly higher compared to that obtained from the I-V measurements. The reason behind this again is related to the spatial variation of barrier height which influences the I-V measurements. C-V measurements on the other hand give the average value of barrier height calculated from the capacitance involving the entire contact area.

The quality of the SI epitaxial layers was evaluated using preferential etching and XRD rocking curve measurement techniques. The studies revealed the high quality of the epitaxial layers which showed the width of the rocking curve peak to be as low as 3.6 arc sec corresponding to the (0008) plane reflection. Theoretical calculations predicted the width of the rocking curve peak for a similar plane to be 2.7 arc sec. However, the presence of structural defects like threading screw dislocations (TSDs), threading edge dislocations (TEDs), and basal plane dislocations (BPDs) were confirmed using KOH etching and optical spectroscopy.

The electrical measurements revealed that the n-type epitaxial layers, owing to their capability of formation of Schottky diodes and less defect interference, are more suitable for detector fabrication. Hence, detectors were fabricated on n-type epitaxial layers and tested for their performance. For the radiation detection measurements mostly 20 and 50 μm superior quality n-type Ni/4H-SiC epitaxial layer SBD were used. For the calibration of our nuclear spectrometer an estimation of the electron hole-pair creation energy (EHP) was done using a method of absolute calibration. A value of ≈7.3 eV electron hole-pair creation energy was calculated and used for the subsequent measurements. Using alpha particle spectrometry, the charge collection efficiency was determined for these detectors for different applied bias voltages using which the minority (hole) carrier diffusion length in these epitaxial layers was found to be ≈18.6 μm.

These detectors readily detected alpha particles with high efficiency even without any applied bias. Under optimized biasing and shaping conditions, an extremely high energy-resolution of ≈0.29% was achieved for 5486 keV alpha particles, without using a collimated source. Careful electronic noise analysis showed that the intrinsic detector resolution to be 10.5 keV for 5486 keV alpha particles. It was also revealed from the noise analysis that the white series noise of the spectrometer increased when the detector was plugged in and the optimized shaping time was found to shift towards 6 μs compared to 2 μs when the detector was not plugged in. These detectors also showed a very high sensitivity towards X-rays and low energy gamma rays. They also exhibited high spatial uniformity of X-ray responsivity. Pulse height spectrum revealed an energy resolution of 2.1% for 59.5 keV peak from a ^{241}Am source which is comparable to that normally obtained from CdZnTe (CZT) detectors.

The subsequent studies followed on investigating the defects that control the ultimate performance of these devices. The 4H-SiC semi-insulating and good quality n-type epitaxial layers were studied using KOH etching and optical microscopy. The n-type samples showed features like comet tails, pits, hillocks, triangular defects, and step bunching, and the SI epitaxial layers showed the presence of carrot defects only. Correlation with EBIC studies on the n-type samples showed the influence of comet tail morphological defects on the current flow through the epilayers. The superior quality n-type epitaxial layers did not show the presence of etch-pits. Other sensitive techniques like TSC and DLTS measurements verified the presence of defects like Ti(h), Ti(c), $Z_{1/2}$, EH_5, Ci1, IL1, EH_6, and EH_7, with a few unidentified ones. Of all the defects, $Z_{1/2}$ defects, which are identified as carbon vacancies and located at 0.67 eV below the conduction band edge, were seen to directly affect the detector properties most. The best detector that was tested was found to have a nominal concentration of $Z_{1/2}$ defects compared to the rest of the detectors. Isochronal annealing studies showed that all the visible defect levels remained quite stable up to an annealing temperature of 800 °C and duration of 30 min.

5. Conclusions

The advent of 4H-SiC epitaxial layers and SI 4H-SiC single crystals as radiation detector materials were thoroughly studied and results were presented in terms of investigations done in our laboratories at UofSC. Our studies revealed the essential factors which regulate the performance of 4H-SiC epitaxial layer based Schottky barrier detectors. We found that the key to the performance of these devices are thin nickel contacts deposited on RCA cleaned surface which resulted in high surface barrier heights. We also observed that different sister samples obtained from the same parent wafer behave differently when compared in terms of barrier heights, ideality factors, and alpha detection resolution. It was also established that the ultimate performance of these detectors was primarily controlled by the type, concentration, and capture cross-section of the intrinsic point defects. Presence of one particular defect related to carbon vacancies, the $Z_{1/2}$ center, was identified as the primary factor behind poor resolution of 4H-SiC epitaxial layer based SBDs.

6. Patents

Schottky barrier detection devices having a 4H-SiC n-type epitaxial layer, US 9,515,211 B2 (2016).

Author Contributions: Investigation, K.C.M., J.W.K., and S.K.C.; Supervision, K.C.M., Writing—Original draft, K.C.M., S.K.C., and J.W.K.; Writing—Review and editing, K.C.M., and S.K.C. All authors have read and agreed to the published version of the manuscript.

Funding: This work was partly supported by the DOE Office of Nuclear Energy's Nuclear Energy University Program (NEUP), Grant No. DE-AC07-051D14517, DE-NE0008662, and by Los Alamos National Laboratory/DOE (Grant #143479). The work was also supported in part by the Advanced Support Program for Innovative Research Excellence-I (ASPIRE-I) of the University of South Carolina (UofSC), Columbia, Grant No. 15530-E404.

Conflicts of Interest: The authors declare no conflicts of interest. The funding sponsors had no role in the design of the study; in the collection, analyses, or interpretation of data; in the writing of the manuscript, and in the decision to publish the results.

References

1. Babcock, R.V.; Chang, H.C. SiC neutron detectors for high-temperature operations, neutron dosimetry. In Proceedings of the Symposium on Neutron Detection, Dosimetry and Standardization, Vienna, Austria, 10–14 December 1962; International Atomic Energy Agency (IAEA): Vienna, Austria, 1962; Volume 1, pp. 613–622.
2. Dulloo, A.R.; Ruddy, F.H.; Seidel, J.G.; Davison, C.; Flinchbaugh, T.; Daubenspeck, T. Simultaneous measurement of neutron and gamma-ray radiation levels from a TRIGA reactor core using silicon carbide semiconductor detectors. *IEEE Trans. Nucl. Sci.* **1999**, *46*, 275–279. [CrossRef]
3. Nava, F.; Vanni, P.; Lanzieri, C.; Canali, C. Epitaxial silicon carbide charge particle detectors. *Nucl. Instrum. Methods Phys. Res. Sect. A Accel. Spectrom. Detect. Assoc. Equip.* **1999**, *437*, 354–358. [CrossRef]
4. Nava, F.; Bertuccio, G.; Cavallini, A.; Vittone, E. Silicon carbide and its use as a radiation detector material. *Meas. Sci. Technol.* **2008**, *19*, 102001. [CrossRef]
5. Puglisi, D.; Bertuccio, G. Silicon carbide microstrip radiation detector. *Micromachines* **2019**, *10*, 835. [CrossRef] [PubMed]
6. Lucas, G.; Pizzagalli, L. Comparison of threshold displacement energies in β-SiC determined by classical potentials and ab initio calculations. *Nucl. Instrum. Methods Phys. Res. Sect. B Beam Interact. Mater. Atoms.* **2005**, *229*, 359–366. [CrossRef]
7. Mandal, K.C.; Krishna, R.M.; Muzykov, P.G.; Das, S.; Sudarshan, T.S. Characterization of semi-insulating 4H silicon carbide for radiation detectors. *IEEE Trans. Nucl. Sci.* **2011**, *58*, 1992–1999. [CrossRef]
8. Ruddy, F.H.; Seidel, J.G.; Chen, H.; Dulloo, A.R.; Ryu, S. High resolution alpha-particle spectrometry using 4H silicon carbide semiconductor detectors. *IEEE Trans. Nucl. Sci.* **2006**, *53*, 1713–1718. [CrossRef]

9. Bertuccio, G.; Casiraghi, R.; Nava, F. Epitaxial silicon carbide for X-ray detection. *IEEE Trans. Nucl. Sci.* **2001**, *48*, 232–233. [CrossRef]
10. Sellin, P.J.; Vaitkus, J. New materials for radiation hard semiconductor detectors. *Nucl. Instrum. Meth. Phys. Res. Sect. A Accel. Spectrom. Detect. Assoc. Equip.* **2006**, *557*, 479–489. [CrossRef]
11. Kern, W. The Evolution of Silicon Wafer Cleaning Technology. *J. Electrochem. Soc.* **1990**, *137*, 1887–1892. [CrossRef]
12. Mandal, K.C.; Muzykov, P.G.; Krishna, R.M.; Hayes, T.; Sudarshan, T.S. Thermally stimulated current and high temperature resistivity measurements of 4H semi-insulating silicon carbide. *Solid State Commun.* **2011**, *151*, 532–535. [CrossRef]
13. Keister, J. Absolute Radiometric Calibration at NSLS. *Synchr. Radiat. News* **2007**, *19*, 14–17. [CrossRef]
14. Muzykov, P.G.; Krishna, R.M.; Mandal, K.C. Temperature dependence of current conduction in semi-insulating 4H-SiC epitaxial layer. *Appl. Phys. Lett.* **2012**, *100*, 032101. [CrossRef]
15. Lampert, M.A.; Mark, P. *Current Injection in Solids*; Academic: New York, NY, USA, 1970.
16. Sze, S.M. *Semiconductor Devices: Physics and Technology*, 2nd ed.; John Wiley & Sons, Inc.: Hoboken, NJ, USA, 2002.
17. Mandal, K.C.; Muzykov, P.G.; Chaudhuri, S.K.; Terry, J.R. Low Energy X-Ray and γ-Ray Detectors Fabricated on n-Type 4H-SiC Epitaxial Layer. *IEEE Trans. Nucl. Sci.* **2013**, *60*, 2888–2893. [CrossRef]
18. Roccaforte, F.; La Via, F.; Raineri, V.; Pierobon, R.; Zanoni, E. Richardson's constant in inhomogeneous silicon carbide Schottky contacts. *J. Appl. Phys.* **2003**, *93*, 9137–9144. [CrossRef]
19. Tung, R.T. Electron transport at metal-semiconductor interfaces: General theory. *Phys. Rev. B* **1992**, *45*, 13509–13523. [CrossRef]
20. Chaudhuri, S.K.; Krishna, R.M.; Zavalla, K.J.; Mandal, K.C. Schottky barrier detectors on 4H-SiC n-type epitaxial layer for alpha particles. *Nucl. Instrum. Meth. Phys. Res. Sect. A Accel. Spectrom. Detect. Assoc. Equip.* **2013**, *701*, 214–220. [CrossRef]
21. Ichikawa, Y.; Ichimura, M.; Kimoto, T.; Kato, M. Passivation of surface recombination at the Si-face of 4H-SiC by acidic solutions. *ECS J. Solid State Sci. Technol.* **2018**, *7*, Q127–Q130. [CrossRef]
22. Nguyen, K.V.; Mannan, M.A.; Mandal, K.C. Improved n-type 4H-SiC Epitaxial Radiation Detectors by Edge Termination. *IEEE Trans. Nucl. Sci.* **2015**, *62*, 3199–3206. [CrossRef]
23. Chen, Y. Defects Structures in Silicon Carbide Bulk Crystals, Epilayers and Devices. Ph.D. Thesis, State University of New York, Stony Brook, New York, NY, USA, May 2008.
24. Mandal, K.C.; Muzykov, P.G.; Chaudhuri, S.K.; Terry, J.R. Assessment of 4H-SiC epitaxial layers and high resistivity bulk crystals for radiation detectors. *Proc. SPIE* **2012**, *8507*, 85070C.
25. Authier, A. *Dynamical Theory of X-ray Diffraction*; Oxford University Press: Oxford, UK, 2004.
26. MacGillavry, C.H.; Rieck, G.D. (Eds.) *International Tables for X-ray Crystallography*; Kynoch Press: Birmingham, England, UK, 1968.
27. Scattering Factors. Available online: http://www.ruppweb.org/new_comp/scattering_factors.htm (accessed on 27 February 2020).
28. Creagh, D. Tables of X-ray absorption corrections and dispersion corrections: The new versus the old. *Nucl. Instrum. Meth. Phys. Res. Sect. A Accel. Spectrom. Detect. Assoc. Equip.* **2013**, *295*, 417–434. [CrossRef]
29. Chaudhuri, S.K.; Zavalla, K.J.; Mandal, K.C. Experimental determination of electron-hole pair creation energy in 4H-SiC epitaxial layer: An absolute calibration approach. *Appl. Phys. Lett.* **2013**, *102*, 031109. [CrossRef]
30. Lo Giudice, A.; Fizzotti, F.; Manfredotti, C.; Vittone, E.; Nava, F. Average energy dissipated by mega-electron-volt hydrogen and helium ions per electron-hole pair generation in 4H-SiC. *Appl. Phys. Lett.* **2005**, *87*, 222105. [CrossRef]
31. Rogalla, M.; Runge, K.; Soldner-Rembold, A. Particle detectors based on semi-insulating silicon carbide. *Nucl. Phys. B Beam Interact. Mater. Atoms.* **1999**, *78*, 516–520. [CrossRef]
32. Lebedev, A.A.; Ivanov, A.M.; Strokan, N.B. Radiation hardness of SiC and hard radiation detectors based on the SiC films. *Fiz. Tekhnika Poluprovodn.* **2004**, *38*, 129–150.
33. Ivanov, A.M.; Kalinina, E.V.; Konstantinov, A.O.; Onushkin, G.A.; Strokan, N.B.; Kholuyanov, G.F.; Hallen, A. High-resolution short range ion detectors based on 4H-SiC films. *Tech. Phys. Lett.* **2004**, *30*, 575–577. [CrossRef]
34. Bertuccio, G.; Casiraghi, R. Study of silicon carbide for X-ray detection and spectroscopy. *IEEE Trans. Nucl. Sci.* **2003**, *50*, 175–185. [CrossRef]
35. Knoll, G.F. *Radiation Detection and Measurement*, 3rd ed.; Wiley: New York, NY, USA, 2000.

36. Chaudhuri, S.K.; Zavalla, K.J.; Mandal, K.C. High Resolution Alpha Particle Detection Using 4H-SiC Epitaxial Layers: Fabrication, Characterization, and Noise Analysis. *Nucl. Instrum. Meth. Phys. Res. Sect. A Accel. Spectrom. Detect. Assoc. Equip.* **2013**, *728*, 97–101. [CrossRef]
37. Mandal, K.C.; Chaudhuri, S.K.; Nguyen, K.V. An overview of application of 4H-SiC n-type epitaxial Schottky barrier detector for high resolution nuclear detection. In Proceedings of the 2013 IEEE Nuclear Science Symposium and Medical Imaging Conference (2013 NSS/MIC), Seoul, Korea, 27 October–2 November 2013.
38. Breese, M.B.H. A theory of ion beam induced charge collection. *J. Appl. Phys.* **1993**, *74*, 3789–3799. [CrossRef]
39. Ziegler, J.F.; Biersack, J.P.; Littmark, U. *The Stopping and Range of Ions in Solids*; Pergamon: Oxford, UK, 1985.
40. Mandal, K.C.; Chaudhuri, S.K.; Nguyen, K.V.; Mannan, M.A. Correlation of Deep Levels with Detector Performance in 4H-SiC Epitaxial Schottky Barrier Alpha Detectors. *IEEE Trans. Nucl. Sci.* **2014**, *61*, 2338–2344. [CrossRef]
41. Ruddy, F.H.; Dulloo, A.R.; Seidel, J.G.; Seshadri, S.; Rowland, L.B. Development of a silicon carbide radiation detector. *IEEE Trans. Nucl. Sci.* **1998**, *45*, 536–541. [CrossRef]
42. Ruddy, F.H.; Dulloo, A.R.; Seidel, J.G.; Palmour, J.W.; Singh, R. The charged particle response of silicon carbide semiconductor radiation detectors. *Nucl. Instrum. Methods Phys. Res. Sect. A Accel. Spectrom. Detect. Assoc. Equip.* **2003**, *505*, 159–162. [CrossRef]
43. Ivanov, A.; Kalinina, E.; Kholuyanov, G.; Strokan, N.; Onushkin, G.; Konstantinov, A.; Hallen, A.; Kuznetsov, A. High Energy Resolution Detectors based on 4H-SiC. In Proceedings of the 5th European Conference Silicon Carbide and Related Materials 2004, Bologna, Italy, 31 August–4 September 2004; Nipoti, R., Poggi, A., Scorzoni, A., Eds.; Trans Tech Publications Inc.: Zurich, Switzerland, 2005; p. 1029.
44. Ruddy, F.H.; Seidel, J.G.; Sellin, P. High-resolution alpha spectrometry with a thin-window silicon carbide semiconductor detector. In Proceedings of the 2009 IEEE Nuclear Science Symposium Conference Record (NSS/MIC), Orlando, FL, USA, 25–31 October 2009; p. 2201.
45. Bartlett, R.J.; Trela, W.J.; Michaud, F.D.; Southworth, S.H.; Alkire, R.W.; Roy, P.; Rothe, R.; Walsh, P.J.; Shinn, N. Characteristics and performance of the Los Alamos VUV beamline at the NSLS. *Nucl. Instrum. Meth. Phys. Res. Sect. A Accel. Spectrom. Detect. Assoc. Equip.* **1988**, *266*, 199–204. [CrossRef]
46. Day, R.H.; Blake, R.L.; Stradling, G.L.; Trela, W.J.; Bartlett, R.J. Los Alamos X-ray characterization facilities for plasma diagnostics. *Proc. SPIE* **1986**, *689*, 208–217.
47. Terry, J.R.; Distel, J.R.; Kippen, R.M.; Schirato, R.; Wallace, M.S. Evaluation of COTS silicon carbide photodiodes for a radiation-hard, low-energy X-ray spectrometer. In Proceedings of the 2011 IEEE Nuclear Science Symposium Conference Record, Valencia, Spain, 23–29 October 2011; Volume NP1.M-236, pp. 485–488.
48. Wilson, M.D.; Cernik, R.; Chen, H.; Hansson, C.; Iniewskic, K.; Jones, L.L.; Seller, P.; Veale, M.C. Small pixel CZT detector for hard X-ray spectroscopy. *Nucl. Instrum. Meth. Phys. Res. Sect. A Accel. Spectrom. Detect. Assoc. Equip.* **2011**, *652*, 158–161. [CrossRef]
49. Chaudhuri, S.K.; Krishna, R.M.; Zavalla, K.J.; Matei, L.; Buliga, V.; Groza, M.; Burger, A.; Mandal, K.C. $Cd_{0.9}Zn_{0.1}Te$ Crystal Growth and Fabrication of Large Volume Single-Polarity Charge Sensing Gamma Detectors. *IEEE Trans. Nucl. Sci.* **2013**, *60*, 2853–2858. [CrossRef]
50. Bertuccio, G.; Pullia, A. A method for the determination of the noise parameters in pre-amplifying systems for semiconductor radiation detectors. *Rev. Sci. Instrum.* **1993**, *64*, 3294–3298. [CrossRef]
51. Mandal, K.C.; Muzykov, P.G.; Krishna, R.M.; Terry, J.R. Characterization of 4H-SiC Epitaxial Layers and High-Resistivity Bulk Crystals for Radiation Detectors. *IEEE Trans. Nucl. Sci.* **2012**, *59*, 1591–1596. [CrossRef]
52. Muzykov, P.G.; Krishna, R.; Das, S.; Hayes, T.; Sudarshan, T.S.; Mandal, K.C. Characterization of 4H silicon carbide using electron beam induced current. *Mater. Lett.* **2011**, *65*, 911–914. [CrossRef]
53. Konstantinov, A.O.; Hallin, C.; Pecz, B.; Kordina, O.; Janzen, E. The mechanism for cubic SiC formation on off-oriented substrates. *J. Cryst. Growth* **1997**, *178*, 495–504. [CrossRef]
54. Neudeck, P.G. Electrical impact of SiC structural crystal defects on high electric field devices. *Mater. Sci. Forum* **2000**, *338–342*, 1161–1166. [CrossRef]
55. Wang, Y.; Ali, G.N.; Mikhov, M.K.; Vaidyanathan, V.; Skrommea, B.J. Correlation between morphological defects, electron beam-induced current imaging, and the electrical properties of 4H-SiC Schottky diodes. *J. Appl. Phys.* **2005**, *97*, 013540. [CrossRef]
56. Muzykov, P.G.; Krishna, R.M.; Mandal, K.C. Characterization of deep levels in n-type and semi-insulating 4H-SiC epitaxial layers by thermally stimulated current spectroscopy. *J. Appl. Phys.* **2012**, *111*, 014910. [CrossRef]

57. Danno, K.; Kimoto, T. Deep level transient spectroscopy on as-grown and electron-irradiated p-type 4H-SiC epilayers. *J. Appl. Phys.* **2007**, *101*, 103704. [CrossRef]
58. Pintilie, I.; Pintilie, L.; Irmscher, K.; Thomas, B. Deep levels in as-grown 4H-SiC epitaxial layers and their correlation with CVD parameters. *Mater. Sci. Forum* **2003**, *433*, 463–466. [CrossRef]
59. Lebedev, A.A. Deep level centers in silicon carbide: A review. *Semiconductors* **1999**, *33*, 107–130. [CrossRef]
60. Huang, Z.C.; Xie, K.; Wie, C.R. A simple and reliable method of thermoelectric effect spectroscopy for semi-insulating III-V semiconductors. *Rev. Sci. Instrum.* **1991**, *62*, 1951–1954. [CrossRef]
61. Zhang, J.; Storasta, L.; Bergman, J.P.; Son, N.T.; Janzén, E. Electrically active defects in n-type 4H–silicon carbide grown in a vertical hot-wall reactor. *J. Appl. Phys.* **2003**, *93*, 4708–4714. [CrossRef]
62. Fang, Z.-Q.; Claflin, B.; Look, D.C.; Polenta, L.; Mitchel, W.C. Thermally stimulated current spectroscopy of high-purity semi-insulating 4H-SiC substrates. *J. Electron. Mater.* **2005**, *34*, 336–340. [CrossRef]
63. Mannan, M.A.; Chaudhuri, S.K.; Nguyen, K.V.; Mandal, K.C. Effect of $Z_{1/2}$, EH_5, and Ci1 deep defects on the performance of n-type 4H-SiC epitaxial layers Schottky detectors: Alpha spectroscopy and deep level transient spectroscopy studies. *J. Appl. Phys.* **2014**, *115*, 224504. [CrossRef]
64. Dalibor, T.; Pensl, G. Electrical properties of the titanium acceptor in silicon carbide. *Phys. Rev. B* **1997**, *55*, 13618–13624. [CrossRef]
65. Dalibor, T.; Pensl, G.; Matsunami, H.; Kimoto, T.; Choyke, W.J.; Schöner, A.; Nordell, A.N. Deep defect centers in silicon carbide monitored with deep level transient spectroscopy. *Phys. Status Solidi* **1997**, *162*, 199–225. [CrossRef]
66. Gelczuk, Ł.; Dabrowska-Szata, M.; Sochacki, M.; Szmidt, J. Characterization of deep electron traps in 4H-SiC Junction Barrier Schottky rectifiers. *Solid State Electron.* **2014**, *94*, 56–60. [CrossRef]
67. Castaldini, A.; Cavallini, A.; Polenta, L.; Nava, F.; Canali, C.; Lanzieri, C. Deep levels in silicon carbide Schottky diodes. *Appl. Surf. Sci.* **2002**, *187*, 248–252. [CrossRef]
68. Storasta, L.; Bergman, J.P.; Janzen, E.; Henry, A.; Lu, J. Deep levels created by low energy electron irradiation in 4H-SiC. *J. Appl. Phys.* **2004**, *96*, 4909–4915. [CrossRef]
69. Kimoto, T.; Danno, K.; Suda, J. Lifetime-killing defects in 4H-SiC epilayers and lifetime control by low-energy electron irradiation. *Phys. Status Solidi* **2008**, *245*, 1327–1336. [CrossRef]
70. Hemmingsson, C.; Son, N.T.; Kordina, O.; Bergman, J.P.; Janzen, E.; Lindstrom, J.L.; Savage, S.; Nordell, N. Deep level defects in electron-irradiated 4H SiC epitaxial layers. *J. Appl. Phys.* **1997**, *81*, 6155–6159. [CrossRef]
71. Alfieri, G.; Kimoto, T. Capacitance spectroscopy study of deep levels in Cl-implanted 4H-SiC. *J. Appl. Phys.* **2012**, *112*, 63717. [CrossRef]
72. Kleppinger, J.W.; Pershin, Y.; Rak, Z.; Mandal, K.C. Investigation on origin of Ru-induced deep-level defects in 4H-SiC epilayer based Schottky diodes by DLTS and theoretical calculations. *Proc. SPIE* **2019**, *11114*, 111140T.
73. Nguyen, K.V.; Mandal, K.C. Ru-induced Deep Levels in Ru/4H-SiC Epilayer Schottky Diodes by Deep Level Transient Spectroscopy. *ECS J. Solid State Sci. Technol.* **2016**, *5*, P3078–P3081. [CrossRef]
74. Beyer, F.C.; Hemmingsson, C.G.; Pedersen, H.; Henry, A.; Isoya, J.; Morishita, N.; Ohshima, T.; Janzen, E. Capacitance transient study of a bistable deep level in e^--irradiated n-type 4H-SiC. *J. Phys. D Appl. Phys.* **2012**, *45*, 45530. [CrossRef]
75. Danno, K.; Kimoto, T. Investigation of deep levels in n-type 4H-SiC epilayers irradiated with low-energy electrons. *J. Appl. Phys.* **2006**, *100*, 113728. [CrossRef]
76. Storasta, L.; Tsuchida, H.; Miyazawa, T.; Ohshima, T. Enhanced annealing of the $Z_{1/2}$ defect in 4H–SiC epilayers. *J. Appl. Phys.* **2008**, *103*, 013705. [CrossRef]
77. Son, N.T.; Trinh, X.T.; Løvlie, L.S.; Svensson, B.G.; Kawahara, K.; Suda, J.; Kimoto, T.; Umeda, T.; Isoya, J.; Makino, T.; et al. Negative-U System of Carbon Vacancy in 4H-SiC. *Phys. Rev. Lett.* **2012**, *109*, 187603. [CrossRef] [PubMed]
78. Nava, F.; Castaldini, A.; Cavallini, A.; Errani, P.; Cindro, V. Radiation Detection Properties of 4H-SiC Schottky Diodes Irradiated Up to 10^{16} n/cm^2 by 1 MeV Neutrons. *IEEE Trans. Nucl. Sci.* **2006**, *53*, 2977–2982. [CrossRef]
79. Gordon, L.; Janotti, A.; Van de Walle, C.G. Defects as qubits in 3C− and 4H−SiC. *Phys. Rev. B* **2015**, *92*, 045208. [CrossRef]
80. Reshanov, S.A.; Pensl, G.; Danno, K.; Kimoto, T.; Hishiki, S.; Ohshima, T.; Itoh, H.; Yan, F.; Devaty, R.P.; Choyke, W.J. Effect of the Schottky barrier height on the detection of midgap levels in 4H-SiC by deep level transient spectroscopy. *J. Appl. Phys.* **2007**, *102*, 113702. [CrossRef]

81. Wong-Leung, J.; Svensson, B.G. Electric field assisted annealing and formation of prominent deep-level defect in ion-implanted n-type 4H-SiC. *Appl. Phys. Lett.* **2008**, *92*, 142105. [CrossRef]
82. Mannan, M.A.; Nguyen, K.V.; Pak, R.O.; Oner, C.; Mandal, K.C. Deep Levels in n-Type 4H-Silicon Carbide Epitaxial Layers Investigated by Deep-Level Transient Spectroscopy and Isochronal Annealing Studies. *IEEE Trans. Nucl. Sci.* **2016**, *63*, 1083–1090. [CrossRef]

micromachines

MDPI

Article

Modeling and Analysis of v_{gs} Characteristics for Upper-Side and Lower-Side Switches at Turn-on Transients for a 1200V/200A Full-SiC Power Module

Maosheng Zhang [1], Na Ren [1], Qing Guo [1,*], Xiangwen Zhu [1], Junming Zhang [1] and Kuang Sheng [1,2]

1 College of Electrical Engineering, Zhejiang University, Hangzhou 310027, China; zhangmaosheng@zju.edu.cn (M.Z.); ren_na@zju.edu.cn (N.R.); zhuxiangwen@zju.edu.cn (X.Z.); zhangjm@zju.edu.cn (J.Z.); shengk@zju.edu.cn (K.S.)

2 State Key Laboratory of Advanced Power Transmission Technology, Beijing 102209, China

* Correspondence: guoqing@zju.edu.cn; Tel.: +86-571-8795-1345

Received: 19 October 2019; Accepted: 16 December 2019; Published: 18 December 2019

Abstract: In this work, a 1200V/200A full-SiC half-bridge power module was fabricated for high-power high-frequency application, and the characteristics of gate-source voltage (v_{gs}) at turn-on transient under different output power was investigated via experiments, modeling, and simulation. Also, the comparison of the v_{gs} characteristics between the upper-side and lower-side was conducted. From experiments, the v_{gs} characteristics show negative spike issue and it becomes severe under higher output power conditions. On the other hand, the upper-side and lower-side show different characteristics, namely, the v_{gs} spike of upper-side is superimposed by a 83.3 MHz high frequency oscillation during the process of v_{gs} being pulled down, while the v_{gs} spike of lower-side contains no oscillation. The mechanisms behind the influence of output power on the v_{gs} spike characteristics and their difference between the upper-side and lower-side were studied via modeling and simulation. Equivalent RLC (resistance-inductance-capacitance) circuit models were proposed and established for the gate driver loops of the upper-side and lower-side based on the internal structure of the power module. With the help of the proposed models, v_{gs} characteristics of the upper-side and lower-side were simulated and compared with the experimental results. The trend of changes in the v_{gs} pulling-down and oscillation amplitude along with the increasing output power from simulation are consistent with that of the experimental results. In addition, different conditions of gate resistance for the SiC power module are compared. Through the comparison between the experiments and simulations, the validity of the proposed equivalent RLC circuit model and the rationality of the analysis about the mechanisms behind the v_{gs} characteristics at turn-on transient for SiC half-bridge power module are confirmed.

Keywords: silicon carbide; power module; negative gate-source voltage spike

1. Introduction

Compared with the counterpart of silicon devices, SiC power semiconductor devices can operate at a higher frequency because the switching time of Si-IGBT (silicon based insulated gate bipolar transistor) ranges from 50 ns to 200 ns while it decreases to 10–20 ns for SiC devices [1–4]. Combined with the parasitic parameters of SiC devices and/or modules, the higher switching frequency and higher allowable negative gate-source voltage as well as the lower threshold voltage for SiC devices can also raise some issues, such as voltage overshoot [5], power loss [6], EMI (electromagnetic interference) emission [7–10], shoot-through or cross-talk fault [11–16], and switching oscillation [17–22]. These problems have brought great challenges for adopting SiC devices and/or modules into the mainstream power conversion application with high frequency.

The switching transient issues for a half-bridge power module has been investigated by many researchers. For example, reference [23] analyzed the switching resonance phenomenon that occurred during the process of fully turn-on and turn-off. Reference [24] investigated the influence of parasitic element of a discrete SiC MOSFET device on the switching performance, especially on drain-source voltage (v_{DS}) characteristics at turn-on transient. Reference [25,26] investigated the effects of increasing load current/output power on Miller plateau voltage and turn-off transient. It is concluded that the output power has great influence on the characteristics of drain current (i_{DS}), drain-source voltage (v_{DS}), and gate-source voltage (v_{gs}) at turn-off transients.

For the v_{gs} characteristics at switching transient, reference [27] investigated the negative gate-source voltage (v_{gs}) spike issue of the upper-side switch and viewed this negative voltage spike as a crosstalk issue. The peers only studied the interaction between the upper-side and lower-side switches, and they put efforts in designing gate driver circuits to alleviate the negative v_{gs} spike issues [28,29]. However, the influence of output power on the v_{gs} characteristics at turn-on transients, and the difference between the v_{gs} characteristics for the upper-side and lower-side switches is not fully studied.

In order to study the influences of output power on the turn-on v_{gs} characteristics for high-power high-frequency application, a 1200V/200A full-SiC high-power module and an inductive load double-pulse test rig were fabricated in this work. The amplitudes of v_{gs} spike and oscillation under different output power conditions were investigated. By building up two different equivalent RLC circuit models of the gate loop path for the upper-side and lower-side, the characteristics of the v_{gs} spike at turn-on transient were compared between upper-side and lower-side; the mechanisms behind these different characteristics were also analyzed with the model.

The paper is organized as follows. Section 2 introduces the experiment setup which includes the developed high-frequency full-SiC power module and a clamped inductive load double pulse test rig. Typical turn-on switching process for a SiC power module is also discussed in this section. Section 3 shows the experimental results of the turn-on v_{gs} characteristics under different output power conditions for both upper-side and lower-side switches. Furthermore, in Section 4, equivalent RLC circuit models of the gate loop path is established for the upper-side and lower-side to analyze the negative v_{gs} spike and oscillation issues in different output power cases. The simulation results are compared to the experimental results. Finally, Section 5 concludes the article.

2. Experiment background

2.1. Introduction of the Developed SiC Power Module

The structure design of the 1200V/200A full-SiC power module with standard package outline is shown in Figure 1a. The diameter of gate wire is 8 mil while other wires inside power module are 14 mil. The footprint of DBC is 34 mm × 25 mm, and the thickness of upper copper, AlN ceramic, and lower copper are 0.3 mm, 0.63 mm, and 0.3 mm, respectively.

The power module has a half-bridge structure topology as shown in Figure 1b, there are four SiC-MOSFET dies (CPM2-1200-0025B, Cree, Inc., Durham, NC, USA) in parallel connection and four SiC-SBD dies (CPW5-1200-Z050B, Cree, Inc., Durham, NC, USA) anti-paralleled with SiC-MOSFETs within each leg of power module.

This high-power high-frequency SiC power module is developed for the boost converter for hybrid electric vehicles. With an increased switching frequency, the volume of inductor in the boost converter can be reduced, which is desirable for downsizing the power control unit for hybrid electric vehicles. In order to acquire an optimized switching performance in high-power high-frequency application, the design of the power module is optimized from two aspects: One is the symmetry of the structure, the other is the common source inductance.

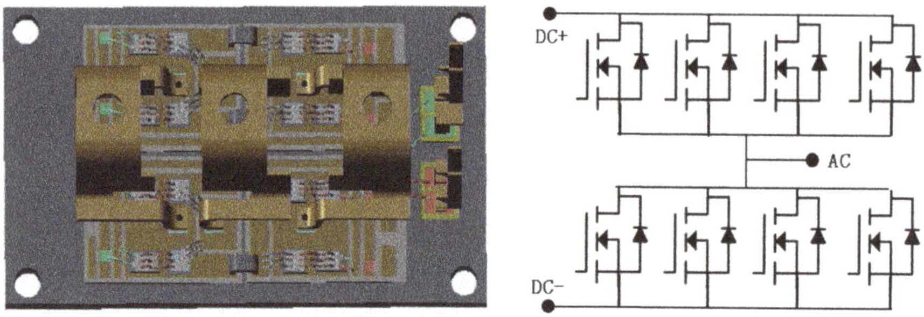

(a) (b)

Figure 1. The structure design of the developed 1200V/200A full-SiC power module: (**a**) Design structure; (**b**) circuit topology.

For the symmetry design, the two parallel-connected DBCs (direct bonding copper) for both upper-side and lower-side are symmetrical, the two parallel-connected pair of SiC-MOSFET and anti-parallel SiC-SBDs for both upper-side and lower-side are also symmetrical. The symmetry of the gate loops between the parallel chips is realized by placing the gate and source routings of the lower-side near the central line of the module as shown in Figure 2a. When compared to some commercial power module (Figure 2b) where the gate and source routings of the lower-side are placed to one side of the power module, our design has an enhanced symmetry for the gate-source routings of the parallel-connected chips. On the other hand, the parasitic inductance of the gate-source loop of our design is also reduced.

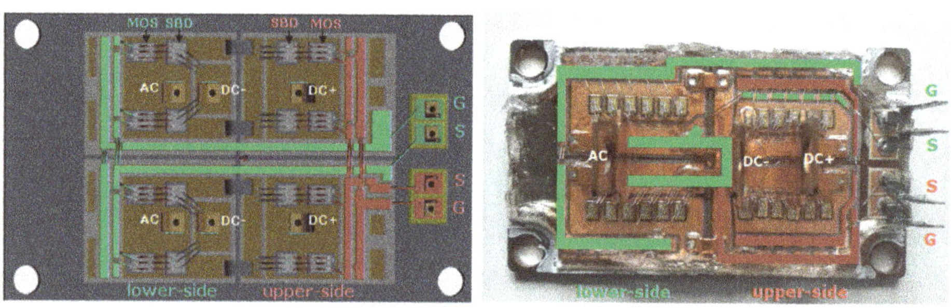

(a) (b)

Figure 2. Comparison of common source paths inside the developed SiC power module with the commercial one. (**a**) Developed module in this work (design structure), (**b**) commercial module (physical structure).

The common source inductance is the parasitic inductance of the common source path of drain routing and gate driver routing inside the power module. Due to its negative feedback effect, a large common source inductance can suppress the change of v_{gs} and slow down the drain current slew rate, which results in a significant increase of switching loss and a decrease of switching frequency [6,24,30–33]. This means that reducing common source inductance must be considered to speed up the switching of the power module.

In order to minimize the common source inductance, the gate-source routing and drain-source routing are separated in our design, as shown in Figure 2a, where the red lines and green lines stand for the gate driver routing inside the power module for upper-side and lower-side, respectively. The drain routing inside the power module for upper-side starts from DC+ terminal, then goes through MOSFET chips, and crosses SBD chips by wire and copper bridge, and finally goes to AC terminal. For lower-side, the drain routing starts from AC terminal, then goes through MOSFET chips, and crosses SBD chips by wire, and finally goes to DC- terminal (see Figure 2a). There is no common path between the gate-source and drain-source routings inside our power module except the power chips themselves, this means that the common source inductance is minimized in our power module. Thus, switching power loss can be reduced and switching speed can be increased.

2.2. Introduction of Double Pulse Testing Platform

The clamped inductive double-pulse test rig is shown in Figure 3a. The DUT is the developed 1200V/200A full-SiC power module. The electrolytic capacitor of FG810K901-1 from VDTCAP (Shenzhen, China) is used as the DC bus capacitor. Two capacitors from KEMET (Fort Lauderdale, FL, USA) is used as decoupling capacitor. Two serial-connected self-fabricated inductors are used as load inductor with a total inductance of 160 µH. The voltage and current signals are acquired by a Teledyne Lecroy HDO6104 1GHz high definition oscilloscope (Teledyne LeCroy, Chestnut Ridge, New York, USA). A Rogowski current waveform transducer CWT miniHF 1B (Power Electronic Measurements Ltd (PEM), Nottingham, UK) with bandwidth 30 MHz is utilized to measure the drain current of the DUT. A passive voltage probe PP026-2 with bandwidth of 500 MHz and a high voltage differential probe HVD3106 with the bandwidth of 120 MHz from Teledyne Lecroy are used to obtain the waveforms of v_{gs} and v_{DS} of DUT respectively.

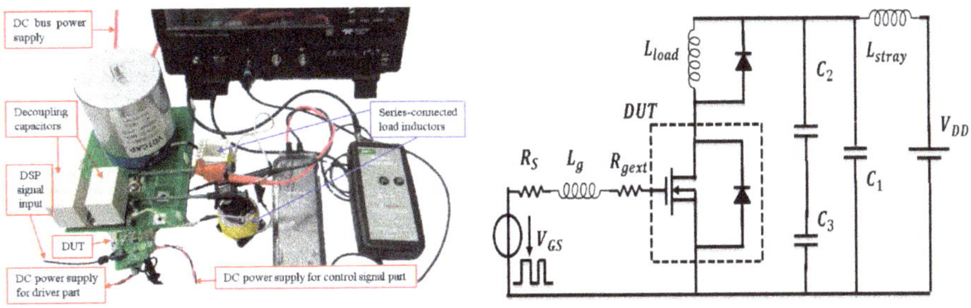

(a) (b)

Figure 3. The clamped inductive load double pulse testing circuit platform: (**a**) Physical picture, (**b**) schematic diagram of circuit.

The values of the components in the testing rig are listed in Table 1. Figure 3b shows the schematic diagram of the testing circuit. The output resistance of the gate driver circuit, R_S, is 0.6 Ω. The gate-source voltage is −2.3 V and 17.5 V for turning off and turning on switches, respectively.

Table 1. The circuit parameters of components of the testing rig.

Components	Names	Value
DC-bus capacitor	C_1	1000 µF
Decoupling capacitor	C_2	3 µF
Decoupling capacitor	C_3	3 µF
Load inductor	L_{load}	160 µH
External gate resistor	R_{gext}	2.4 Ω

2.3. Turn-on Switching Process of SiC Power Module

The typical turn-on switching transient of SiC power module can be divided by four intervals as shown in Figure 4.

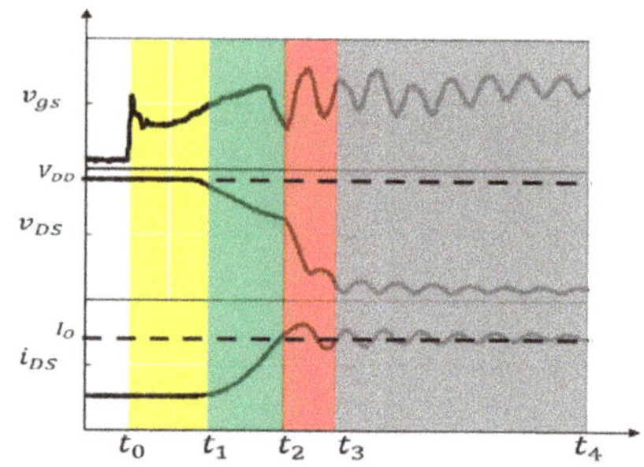

Figure 4. Typical diagram of turn-on switching transient for SiC power module.

- Interval 1: From t_0 to t_1

The yellow region from t_0 to t_1 is the turn-on delay interval, τ_D (on). In this interval the gate voltage charges up the input capacitance of SiC-MOSFETs, the i_{DS} keeps the off-state value until the v_{gs} reaches the threshold voltage, the v_{DS} reduces 10% $\times V_{DD}$ when approaching the end of this period.

- Interval 2: From t_1 to t_2

The green region from t_1 to t_2 is the turn-on interval 1, τ_1 (on). In this interval, the gate voltage continues to charge up the input capacitor of SiC-MOSFETs, the v_{DS} continues to reduce while the i_{DS} rose to I_o at the end of this period.

- Interval 3: From t_2 to t_3

The red region from t_2 to t_3 is the turn-on interval 2, τ_2 (on). During the τ_2 (on), the gate voltage continues to charge up the input capacitor of SiC-MOSFETs, the i_{DS} continues to change, and the v_{DS} reduces to $V_{DS(on)}$ with a higher dv_{DS}/dt in this period. The anti-parallel SiC-SBDs begin to regain reverse blocking capability, the rise in voltage across the anti-parallel SiC-SBD causes the SiC-MOSFET voltage to fall rapidly.

- Interval 4: From t_3 to t_4

The gray region from t_3 to t_4 is used to indicate that the power module is fully turned on, and the time t_4 is at any moment after t_3 during turn-on interval. During the fully turn-on interval, all v_{gs}, v_{DS}, i_{DS} characteristics are kept in oscillation state with their respective frequencies which are determined by the parasitic parameters in the circuit. Due to the different parasitic parameters in drain circuit and gate driver loop, the oscillation frequency of v_{gs} was different from that of v_{DS} and i_{DS}.

During the period from t_1 to t_3, the increasing drain current i_{DS} makes v_{DS} fall. When i_{DS} is less than I_O the freewheeling diodes are forced to conduct current. If the slew rate of v_{DS} is too high before i_{DS} rising to I_o, the v_{DS} will be limited to a voltage platform. This means the slew rate of v_{DS} is not too high in this sub-interval of τ_1 (on). When i_{DS} is increased to be higher than I_O, the rise in blocking voltage across the anti-parallel SiC-SBD causes the SiC-MOSFET voltage to fall rapidly. This means the slew rate of v_{DS} will be higher in this sub-interval of τ_2 (on). Considering the coupling effect of gate-drain capacitor C_{gd} of SiC-MOSFETs inside the power module, the waveform of v_{gs} is affected by the dv_{DS}/dt of drain loop in this period. Meanwhile, the dv_{DS}/dt of drain circuit has greater impacts on the waveform of v_{gs} of gate driver loop during turn-on interval τ_2 (on) rather than τ_1 (on). It will be discussed in the following section.

3. Experimental Results and Comparison

In order to confirm whether our design can reduce the switching time of the power module, our module and one commercial SiC power module (Figure 2b) were tested based on the built double pulse test platform and gate driver circuit. Their transient waveforms are compared in Figure 5.

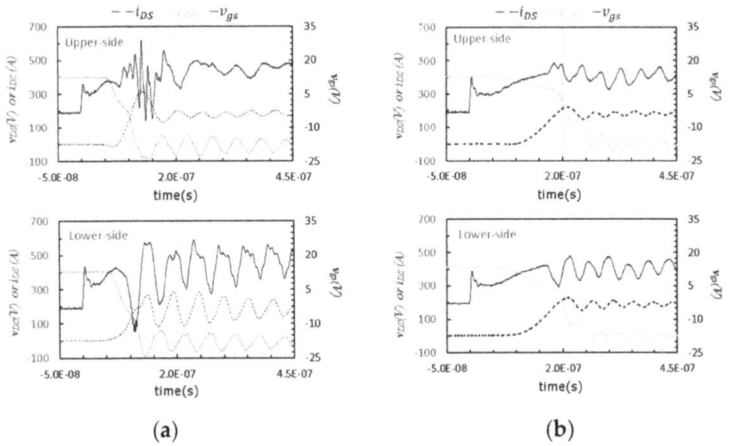

Figure 5. Comparison of transient waveforms between the module developed in this work and one commercial SiC power module (V_{DD} = 400 V, I_O = 200 A). (**a**) Developed SiC power module in this work; (**b**) commercial SiC power module.

As shown in Figure 5, with the same gate driver and double pulse testing platform, the turn-on process of the developed SiC power module was faster than that of the commercial SiC power module. The turn-on times of these two SiC power modules are listed in Table 2. The total turn-on time of the developed module was 110 ns for the upper-side and 117.6 ns for the lower-side. When compared to that of the commercial module, the turn-on time was reduced by 56.4% and 52.0% for the upper-side and lower-side, respectively. Thus, the developed power module is suitable for high frequency application. However, when the turn-on process of the developed power module was speeded up, a severe oscillation and negative voltage spike was observed from the v_{gs} waveforms during the turn-on transient even though the common source inductance was minimized. Therefore, to optimize the high-power high-frequency SiC power module, it is necessary to study the v_{gs} characteristic and find proper designs to address the oscillation and negative voltage spike issues.

Table 2. Comparison of turn-on time for SiC power module.

SiC Module	Commercial Module			Developed Module		
	T_{don} (ns)	T_r (ns)	T_{total} (ns)	T_{don} (ns)	T_r (ns)	T_{total} (ns)
Upper-side	122.8	129.6	252.4	61.2	48.8	110.0
Lower-side	112.0	133.2	245.2	64.8	52.8	117.6

First of all, the factors that take effects on the v_{gs} characteristic will be discussed in the following. It is well known that the increased V_{DD} and I_O affects the slew rate of v_{DS}. Due to the coupling effects of the drain-gate capacitors, the v_{gs} characteristic will change along with the rise of V_{DD} and I_O. In order to investigate thoroughly the influences of V_{DD} and I_O (i.e., output power) on v_{gs} characteristics at turn-on transient for the developed high frequency SiC power module, and clarify the difference between the v_{gs} characteristics of upper-side and lower-side of the SiC power module, some experiments were conducted. The conditions of these experiments are listed in Table 3. As shown in Table 3, the output power of SiC power module in experiment is increased gradually from case_1 to case_3.

Table 3. V_{DD} and I_O parameters for the three cases.

No.	V_{DD} (V)	I_O (A)
Case_1	200	100
Case_2	400	100
Case_3	400	200

In the experiment of case_1, the switching waveforms of v_{gs} were almost normal for both upper-side and lower-side though there was minor difference between them as shown in Figure 6. The v_{gs} of upper-side was pulled down by 12.76 V during the $\tau_2(on)$ interval (as shown in Figure 4) while the lower-side's was pulled down by 8.8 V during the $\tau_1(on)$ interval. The v_{gs} spikes for both sides were kept positive during the turn-on transient, negative v_{gs} spike was absent from the waveforms for both upper-side and lower-side of the SiC power module.

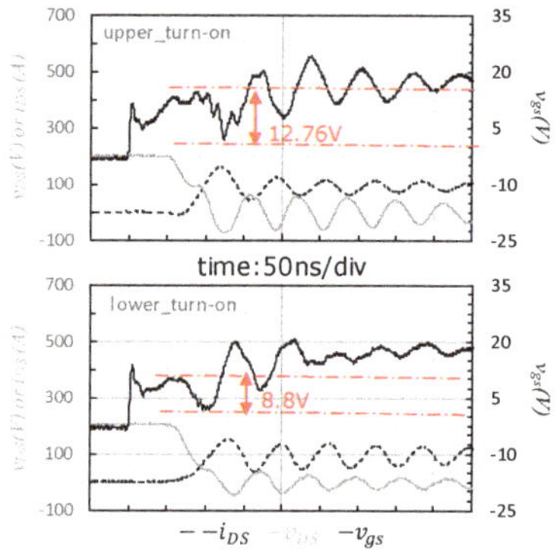

Figure 6. Turn-on transient waveform of case_1.

As output power of the power module rose, the v_{gs} spike started to show different characteristics between the upper-side and lower-side. As shown in Figure 7, the v_{gs} spike of the upper-side was pulled down to an excessively negative voltage (−9.94 V) and accompanied by a significant oscillation with 83.3 MHz frequency during the τ_2 (on) interval. On the other hand, the v_{gs} spike of the lower-side was only pulled down to a negative voltage slightly, and no oscillation was observed from the waveform.

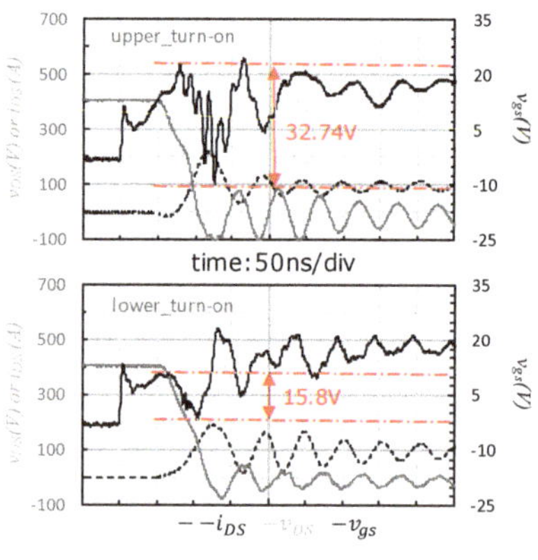

Figure 7. Turn-on transient waveform of case_2.

As we all know, the oscillation is typically a high frequency signal which put stringent requirements on the bandwidth of probes and the contact quality. In our testing experiment, the bandwidth of the voltage probe was 500 MHz, which is five times the oscillation frequency (83.3 MHz); it is also far greater than the minimum bandwidth requirement for fetching this kind of waveform. In order to eliminate the suspicion that the oscillation was a false waveform, we re-examined and adjusted the connection and contact between the oscilloscope probe and the test point, finding out that oscillation still exists.

For the experiment of case_3, the transient waveforms are shown in Figure 8. The v_{gs} spike of the upper-side oscillated more seriously than case_1 and case_2. Both the excessively positive and negative voltage spikes were observed from v_{gs} characteristics of the upper-side, which resulted from the serious oscillation. On the other hand, the v_{gs} of the lower-side was pulled down to a lower negative voltage (−13.7 V), and no oscillation was observed from the waveform.

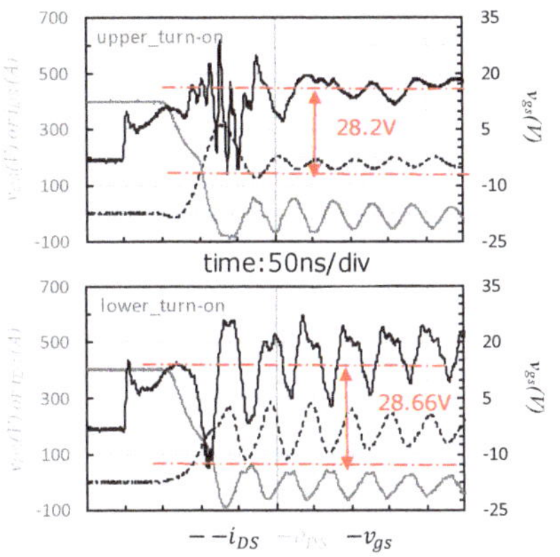

Figure 8. Turn-on transient waveform of case_3.

From the experimental results, it is found that the increase of output power of the SiC power module amplifies the difference of v_{gs} characteristics between upper-side and lower-side. When the output power is relatively low, such as that in case_1, no negative v_{gs} spike appeared; the characteristics of v_{gs} in upper-side are almost the same as that of lower-side. As the output power rises, the negative voltage spike issue of v_{gs} becomes more serious. On the other hand, the difference between the v_{gs} characteristics of the upper-side and lower-side appears. Firstly, the pulling down process of v_{gs} for the upper-side is occurred in the $\tau_2(on)$ interval or the whole period of v_{DS} falling, while that of the lower-side always appears in $\tau_1(on)$ interval. Secondly, the process of v_{gs} pulling down for the upper-side is superimposed by a high frequency oscillation while the lower-side's is absent from oscillation. Thirdly, the oscillation becomes more serious in upper-side and the negative v_{gs} spike becomes lower in lower-side with an increasing output power (from case_1 to case_3).

The turn-on transient waveforms of v_{gs} under different output power conditions are compared in Figure 9, where Figure 9a shows the results of the upper-side and Figure 9b shows the results of the lower-side. The detailed information related to the v_{gs} pulling down process are extracted from Figure 9 and summarized in Table 4, such as amplitude of oscillation, voltage of v_{gs} starts to pull down, the lowest v_{gs} spike and the pull-down amplitude.

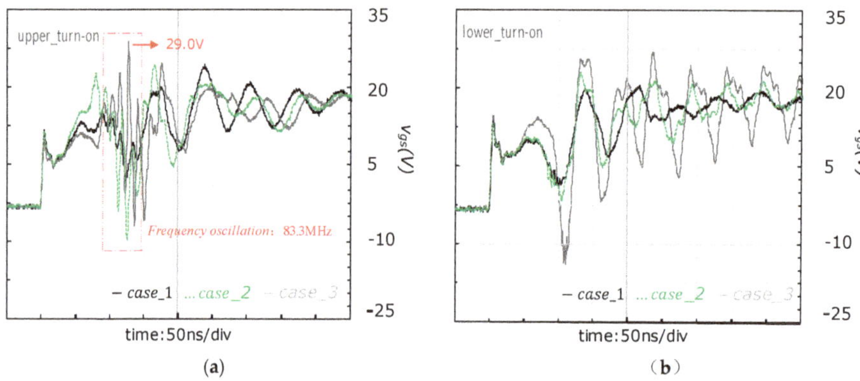

Figure 9. Comparison of turn-on transient waveform of v_{gs} under different output power conditions: (**a**) Turn-on v_{gs} of the upper-side; (**b**) turn-on v_{gs} of the lower-side.

Table 4. Comparison of trainset waveform of v_{gs} at turn-on transient.

Case No.	Location	v_{gs} Value starts to Pull-Down (V)	Amplitude of Oscillation	Lowest v_{gs} Spike (V)	Pull-Down Amplitude (V)
Case_1	Upper-side	14.48	no	1.72	12.76
Case_2	Upper-side	22.80	medium	−9.94	32.74
Case_3	Upper-side	20.44	serious	−7.76	28.20
Case_1	Lower-side	10.32	no	1.52	8.80
Case_2	Lower-side	10.80	no	−4.28	15.08
Case_3	Lower-side	14.96	no	−13.70	28.66

As shown in Table 4, with an increased output power, the lowest v_{gs} spike continues to decrease, and the amplitude of the v_{gs} pulling down rises significantly for the lower-side. On the other hand, for the upper-side, the pulling down amplitude of v_{gs} in case_3 is smaller than that of case_2. This is because the oscillation of v_{gs} in case_3 is more serious than that of case_2 and the highest v_{gs} spike is up to 29.0 V. These differences of the v_{gs} characteristics between the upper-side and the lower-side is probably attributed to the different dv_{DS}/dt of drain loop for the upper-side and lower-side.

As analyzed in reference [26], the parameters such as gate resistance, gate loop inductance, input capacitance C_{iss}, and positive gate-source voltage V_{gs} take effects on the v_{gs} characteristics. In our experiment, the gate driver and testing circuit are the same. This means that the different characteristics of v_{gs} spike between the upper-side and lower-side could be correlated to the different gate-source paths and coupling effects of the dv_{DS}/dt from drain loop to the gate driver loop in the SiC power module.

Meanwhile, we must notice the load current and bus voltage (output power) will influence the switching performance [25]. In other words, even for the same power device, coupling effects between drain loop and gate driver loop by dv_{DS}/dt could be different under different operation conditions. Namely, the different slew rate of v_{DS} during the turn-on transient could bring different extent of coupling influence.

From the experimental results in Figures 6–8, the slew rates of v_{DS} at turn-on switching transient can be extracted. The experimental results of dv_{DS}/dt are summarized in Table 5. It is shown that the slew rate of v_{DS} of the upper-side is slightly larger than that of the lower-side, which means the coupling effect of drain loop to gate loop in the upper-side is more significant than that of the lower-side during turn-on transient.

Table 5. Comparison of dv_{DS}/dt between upper and lower sides.

Conditions	Upper-Side		Lower-Side	
Case No.	dv_{DS}/dt **(V/ns)**	$i_{D \to G}$ **(A)**	dv_{DS}/dt **(V/ns)**	$i_{D \to G}$ **(A)**
Case_1	3.92	0.47	2.91	0.35
Case_2	14.25	1.71	6.24	0.75
Case_3	15.4	1.85	11.13	1.34

4. Modeling and Simulation

4.1. Equivalent Model of Gate-Source Path of SiC Power Module

The equivalent circuit model of the gate loop is obtained based on the analysis of the actual physical structure of the power module and the output stage of the gate driver circuit. For the standard package outline of the half-bridge module, the internal structure of the upper-side is different from that of the lower-side. The routings of drain circuit and gate circuit inside the power module are shown in Figure 10a,b, respectively. Since all the gate input terminals are placed to the outer edge area of the upper-side of the power module, the gate routings of the lower-side must pass through the upper-side before connecting to the power chips of the lower-side.

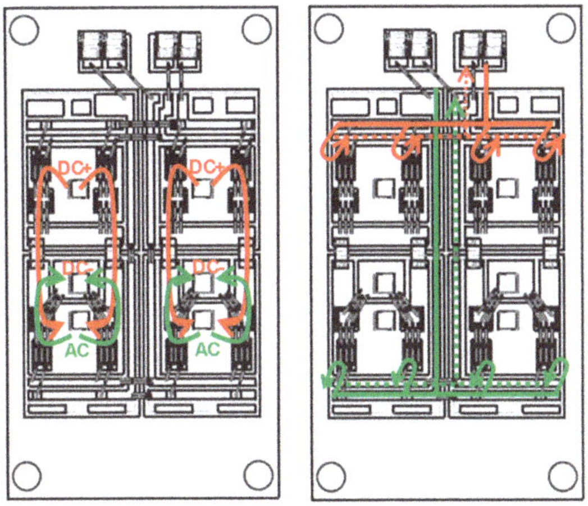

(**a**) (**b**)

Figure 10. Circuit routings inside the power module (red lines for upper-side; green line for lower-side). (**a**) Drain circuit routings, (**b**) gate loop routings.

Compared with gate-source path of the upper-side, the gate-source path of the lower-side is much longer, the input signal must pass through the whole power module before connecting to the power chips (see Figure 2a, Figure 10). As a result, the location of the lumped parasitic inductance of the gate loop in upper-side is different from that of lower-side in our equivalent circuit models at turn-on transient. That is to say, the topology of the equivalent circuit model of the gate driver loop is different between upper-side and lower-side for the developed half-bridge module.

In the previous analysis of the experimental results, we found that the output power of power module affects the characteristics of v_{gs}. As the common source inductance is minimized in our design, the effect of common source inductance on turn-on transient can be ignored. We put emphasis on the coupling effect of drain circuit to gate driver loop, and it will be discussed in the following.

Based on the internal structure of the power module in Figure 10, the equivalent RLC circuit models of the gate driver loop in turn-on transient for the upper-side and lower-side are established and shown in Figure 11a, b, respectively. The coupling effect is equivalent to a short-time current source. As this short-time current source is an external factor for the gate driver loop, it is parallel-connected to the equivalent circuit in the model. With the help of the models, v_{gs} characteristics at the turn-on transient can be simulated.

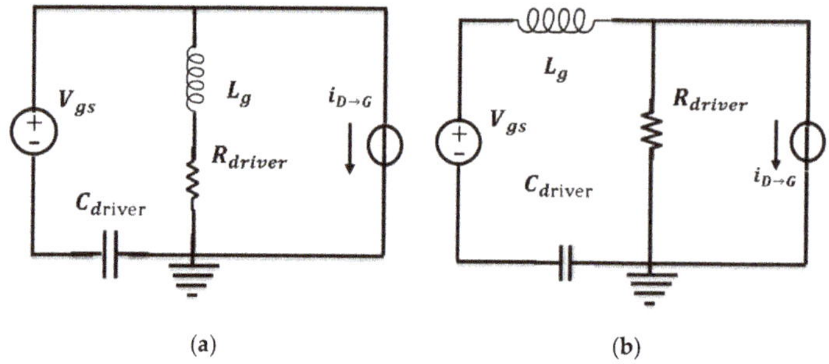

Figure 11. The equivalent circuit models of gate loop path at turn-on transient: (**a**) Upper-side, (**b**) lower-side.

In Figure 11, L_g stands for the total stray inductance of gate driver loop, it includes the gate-source routings inside module, L_{gs}, and the stray inductance of output paths of gate driver circuit, L_s, which can be expressed by Equation (1),

$$L_g = L_{gs} + L_S \tag{1}$$

The R_{driver} is the resistance of the whole gate driver loop, it includes the stray resistance of gate-source routings, R_{gs}, the external gate resistance R_{gext} and internal resistance of SiC-MOSFET, R_{gint}, and the output resistance of gate driver circuit, R_s, which can be expressed by Equation (2),

$$R_{driver} = R_{gs} + R_{gext} + R_{gint} + R_S \tag{2}$$

The C_{driver} is the parasitic capacitance of the total routings of the gate driver loop. The parasitic parameters can be extracted by Q3D software and they are listed in Table 6.

Table 6. Parasitic parameters extracted by Q3D software.

Routings	Symbol	Upper-Side	Lower-Side
Inductance of gate-source routings (nH)	L_{gs}	31.50	61.49
Inductance of output path of gate driver circuit (nH)	L_s	5.00	5.00
Resistance of gate-source routings (Ω)	R_{gs}	0.15	0.31
Resistance of output paths of gate driver circuit (Ω)	R_S	0.60	0.60
Internal resistance of power chips (Ω)	R_{gint}	1.00	1.00
Capacitance of gate driver loop (pF)	C_{driver}	1.00	1.00

The value of the equivalent current source stands for the extent of this influence, which is designated as $i_{D\to G}$. The higher the slew rate of v_{DS}, the higher the $i_{D\to G}$. The $i_{D\to G}$ is given by Equation (3),

$$i_{D\to G} = C_{gd}\frac{dv_{DS}}{dt} \tag{3}$$

The value of $i_{D\to G}$ is determined by C_{gd} and $\frac{dv_{DS}}{dt}$. The capacitance C_{gd} of power chips increases sharply as the voltage v_{DS} decreases at the turn-on transient, and the value of $i_{D\to G}$ is also increased. Therefore, the coupling effect between the drain loop and gate driver loop is enhanced significantly due to the sharply rising C_{gd} at turn-on transient.

If the increase in output power is equivalent to a rise of $\frac{dv_{DS}}{dt}$, $i_{D\to G}$ is getting higher at an increased output power. Thus, the current $i_{D\to G}$ can reflect the extent of influence of output power on gate driver loop in our equivalent circuit model. The equivalent current $i_{D\to G}$ at the turn-on transient for the experiment can be calculated by Equation (3) and they are summarized in Table 5.

Based on the RLC circuit model, the extracted parameters of the power module and gate driver circuit, as well as the calculated $i_{D\to G}$ from the experimental results, the generation mechanism of the characteristics of v_{gs} voltage spike for full-SiC power module at turn-on transient can be studied quantitatively by LTspice software. The respective values of $i_{D\to G}$ for case_1, case_2, and case_3 in simulation are the same with those in experiment (Table 5), while the values of the parasitic parameters used in simulation are from Table 6.

For the upper-side, V_{gs} is set to 15 V, the current source $i_{D\to G}$ starts to output pulse at time of 5 ns and it lasts a time interval of 7 ns in the simulation, the value of $i_{D\to G}$ is as listed in Table 5. The simulation results for the three cases as studied by the experiments in Section 3 are shown in Figure 12.

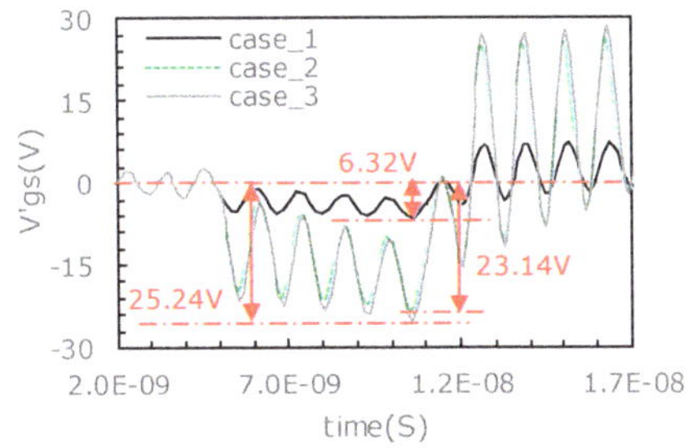

Figure 12. Simulation results of v'_{gs} voltage spike for upper-side.

As the output power increases, the $i_{D\to G}$ rises from 0.47 A to 1.85 A accordingly, the gate-source voltage minus the DC voltage bias (v'_{gs}) is pulled down to a lower value. At the meantime, the gate source voltage is accompanied with a high frequency oscillation. The oscillation frequency is 83.3 MHz, which is the same as that of the experimental results. The oscillation frequency doesn't vary with the $i_{D\to G}$ values as it is only related to the parasitic parameters of both the gate driver circuit and gate-source routings inside the power module. When the equivalent current source starts to output pulse the v'_{gs} tumbles, and the v'_{gs} rebounds immediately when the current source output is terminated.

For the lower-side, the V_{gs} is set to 15 V, the equivalent current source of $i_{D\to G}$ starts to output pulse at time of 6 ns and it lasts a time interval of 5 ns in the simulation. The simulation results for the three cases as studied by the experiments in Section 3 are shown in Figure 13. As the output power increases, the $i_{D\to G}$ rises from 0.35 A to 1.34 A accordingly, v'_{gs} is pulled down to a lower value. There is no high frequency oscillation observed.

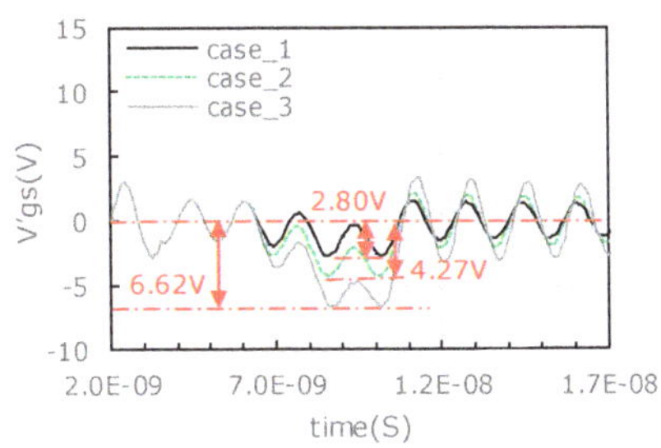

Figure 13. Simulation results of v'_{gs} voltage spike for lower-side.

The characteristics of v'_{gs} pulling down in simulation are compared with the experimental results (as shown in Figure 9). The amplitude of v_{gs} pulling-down and oscillation for the upper-side and lower-side in different cases are listed in Table 7. For the characteristics of high frequency oscillation, the simulation is in good agreement with the experimental results. For the characteristics of the pulling down amplitude of v_{gs} spike, the simulation is almost in agreement with the experimental results except the upper-side.

Table 7. Comparison of characteristics for simulation and experimental results at turn-on transient.

Characteristics of v_{gs} Pulling Down	Pulling Down Amplitude of v_{gs}		Amplitude of High Frequency Oscillation	
	Experiment	Simulation	Experiment	Simulation
Case_1_upper-side	12.76	6.32	tiny	tiny
Case_2_upper-side	32.74	23.14	moderate	moderate
Case_3_upper-side	28.20	25.24	serious	serious
Case_1_lower-side	8.80	2.80	no	no
Case_2_lower-side	15.08	4.27	no	no
Case_3_lower-side	28.66	6.62	no	no

The pulling down amplitude of v_{gs} spike of the lower-side increases as the output power rises in both simulation and experiment, but the pulling down amplitude of v_{gs} of the upper-side in simulation is different from the experiment. In simulation, the pulling down amplitude of v_{gs} rises as output increases, but in experimental results the largest pulling down amplitude of v_{gs} occurred in case_2 rather than case_3. This abnormal phenomenon in the experiment is mainly attributed to the high frequency oscillation on the pulling down waveform of v_{gs}. As shown in Figure 9a, there is an excessively positive voltage spike of 29.0 V in case_3 during the high frequency oscillation process, while the high frequency oscillation doesn't bring the voltage spike back to a lower point as that in case_2.

Overall, the characteristics of v_{gs} spike in simulation almost coincide with that of the experiment results. This proves the rationality of our modeling and analysis about the generation mechanism of v_{gs} voltage spike characteristics for SiC power module.

As depicted previously, the negative v_{gs} voltage spike is correlated to the slew rate of v_{DS} and the resistance of the gate driver loop. On one hand, the increased gate resistance can reduce dv_{DS}/dt, and decrease the coupling between the drain loop and gate driver loop due to a lower $i_{D \to G}$ at a slower slew rate of v_{DS}. On the other hand, the larger the gate resistance of the gate driver loop, the smaller the gate current and voltage spike, and less serious the oscillation at turn-on switching transient. Although the coupling effects are different between the upper-side and lower-side, both the negative v_{gs} spike and high frequency oscillation could shrink as the resistance of the gate driver circuit rises. Another experiment and simulation with a higher external gate resistance are carried out to further verify our RLC circuit model and analysis.

4.2. Verification for the Proposed RLC Circuit Model and Analysis

A resistor of 5 Ω instead of the previous external gate resistor of 2.4 Ω is used in the new experiment. The case_3 is selected for the case study as the output power is highest and the negative spike/oscillation is the most significant in this case. The experimental results are shown in Figure 14.

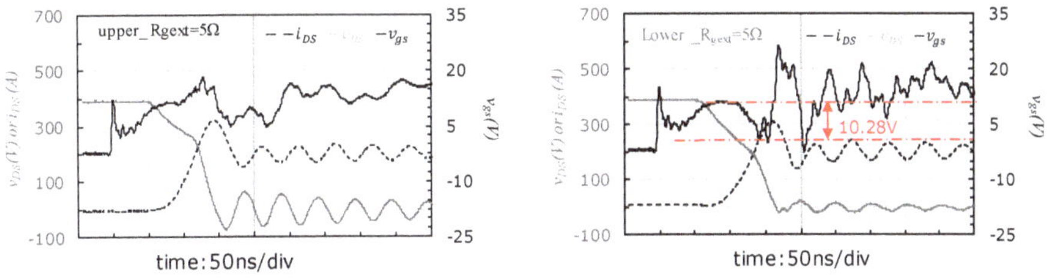

(**a**) (**b**)

Figure 14. Transient waveform at the condition for output power of case_3 and the external gate resistor equal to 5 Ω: (**a**) Upper-side, (**b**) lower-side.

There is neither voltage pulling down nor high frequency oscillation observed for the upper-side (Figure 14a), but a pulling down effect and a spike of v_{gs} is observed for the lower-side (Figure 14b). When the external gate resistance is increased from 2.4 Ω to 5 Ω, the negative spike of v_{gs} in lower-side is decreased from −13.7 V (in Figure 8 and Table 4) to −2.5 V (Figure 14b). Accordingly, the amplitude of v_{gs} pulling down is reduced from 28.66 V (in Figure 8 and Table 4) to 10.28 V (Figure 14b).

The experimental results show that the relatively larger external gate resistor can damp the high frequency oscillation in the upper-side and weaken the amplitude of pulling down of v_{gs} at turn-on transient in SiC power module.

The slew rate of v_{DS} at the turn-on transient are extracted from the experiments and listed in Table 8. The introduced extra current of $i_{D\rightarrow G}$ is calculated with Equation (3) and listed in Table 8 as well.

Table 8. Comparison of dv_{DS}/dt between the cases with external gate resistance of 5 Ω and 2.4 Ω.

Conditions	Upper-Side		Lower-Side	
R_{gext}	dv_{DS}/dt (V/ns)	$i_{D\rightarrow G}$ (A)	dv_{DS}/dt (V/ns)	$i_{D\rightarrow G}$ (A)
2.4 Ω	15.4	1.85	11.13	1.34
5 Ω	3.75	0.45	2.78	0.33

Compared with that of 2.4 Ω, the slew rate of v_{DS} in the condition of gate resistor of 5 Ω reduces significantly; accordingly, the equivalent current $i_{D\rightarrow G}$ from drain loop to gate driver loop due to the coupling effect at the turn-on transient also decreases dramatically.

With the proposed equivalent RLC circuit models shown in Figure 11 and Equations (1)–(3), simulation study with an external gate resistor of 5 Ω is carried out. In this simulation all the parameters are the same as that of the previous simulation except the gate resistance. The simulation results under different external gate resistance conditions are compared in Figure 15. For the upper-side, the high frequency oscillation is alleviated by the higher gate resistance (5 Ω vs. 2.4 Ω), and the amplitude of v_{gs} tumbling is reduced from 25.24 V to 6.92 V (72.5% lower). For the lower-side, the amplitude of v_{gs} tumbling is reduced by ~50% from 6.62 V to 3.25 V.

The characteristics from simulation results (in Figure 15) are extracted and compared with those from the experimental results in Figure 8, 14. The compared characteristics contains the amplitude of v_{gs} pulling down and oscillation. The comparison results are listed in Table 9.

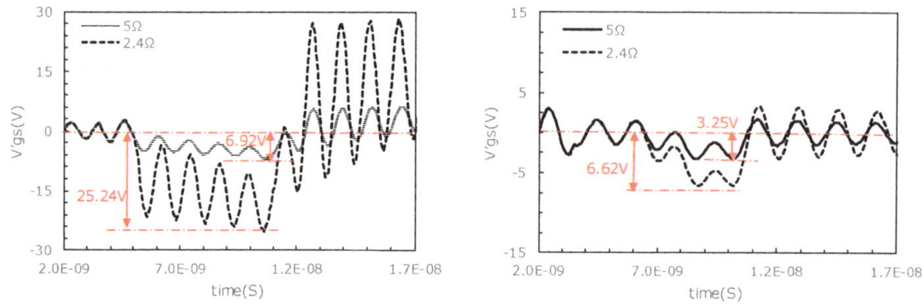

(a) (b)

Figure 15. Simulation results of v_{gs} voltage spike for R_{gext} = 5 Ω and R_{gext} = 2.4 Ω: (**a**) Upper-side, (**b**) lower-side.

Table 9. Comparison of v_{gs} characteristics between experiment and simulation.

Characteristics of v_{gs} Pulling Down	Pulling Down Amplitude of v_{gs} (V)		Amplitude of High Frequency Oscillation	
	Experiment	Simulation	Experiment	Simulation
R_{gext} = 2.4 Ω_upper-side	28.20	25.24	serious	serious
R_{gext} = 5 Ω_upper-side	0	6.92	no	no
R_{gext} = 2.4 Ω_upper-side	28.66	6.62	no	no
R_{gext} = 5 Ω_upper-side	10.28	3.25	no	no

As the gate resistance rises, the characteristics of high frequency oscillation in the upper-side disappears for both experiment and simulation, the amplitude of the pulling down of v_{gs} in the upper-side is lowered significantly for both experiment and simulation. The pulling down amplitude in the experiment reduces from 28.20 V to zero while it decreases from 25.24 V to 6.92 V in the simulation. The amplitude of the pulling down of v_{gs} in the lower-side is also reduced significantly for both experiment and simulation, which reduces from 28.66 V to 10.28 V in the experiment and decreases from 6.62 V to 3.25 V in the simulation.

The specific amplitudes of the pulling down of v_{gs} between the simulation and experiment are different, the error could come from the read of dv_{DS}/dt and the parameter extraction by Q3D software. As shown in Table 9, the trend of changes in the amplitude of v_{gs} pulling down and oscillation along with the increase of output power in simulation generally agrees with those of the experimental results. This further confirms the validity of the proposed equivalent RLC circuit model and the rationality of the analysis about the mechanisms behind the v_{gs} characteristics at turn-on transient for the SiC half-bridge power module.

As guided by the proposed model, the gate driver design must be considered together with the power module design for obtaining an optimized switching performance for the high-power high-frequency SiC power module. For a fabricated power module, we can set the parameters in the model related to the gate driver circuit, such as the L_S in Equation (1), the R_{gext} and R_S in Equation (2). Then the v_{gs} characteristic can be simulated and optimized by tuning the parameters. Thus, a proper design of the gate driver circuit matched with the power module design and output power level can be obtained in a short design cycle.

5. Conclusions

In this paper, a 1200V/200A full-SiC half-bridge power module was fabricated for high-power high-frequency application. The power module is designed with a symmetrical structure and minimized common source inductance to pursue a faster switching. However, severe oscillation and negative voltage spike issues are observed from the v_{gs} waveforms during the turn-on transient, especially at higher output power level.

The characteristics of v_{gs} at turn-on transient under different output power were investigated and their comparison between the upper-side and lower-side was conducted. From experiments, the v_{gs} characteristics show negative spike issue and it becomes severe under higher output power conditions. On the other hand, the upper-side and lower-side show different characteristics, namely, the v_{gs} spike of upper-side is superimposed by a 83.3 MHz high frequency oscillation during the process of v_{gs} being pulled down, while the v_{gs} spike of lower-side contains no oscillation.

The mechanisms behind the influence of output power on the v_{gs} characteristics and the difference of v_{gs} characteristics between upper-side and lower-side were studied via modeling and simulation. Equivalent RLC circuit models were proposed and established for the gate driver loop based on the internal structure of the power module. In the models, the coupling effects between drain circuit and gate driver loop is considered and equalized by a current source ($i_{D\rightarrow G}$). The value of the equivalent current source is determined by gate-drain capacitance C_{gd} and dv_{DS}/dt. As the increase of output power will contribute to a higher dv_{DS}/dt, $i_{D\rightarrow G}$ in the model is increased, i.e., the coupling effect between the drain circuit and gate driver loop is enhanced. Thus, the negative v_{gs} spike issue becomes severe in higher output power conditions. On the other hand, when comparing the upper-side and lower-side, the models are different for them as the gate-source path routings are different. Thus, they show different v_{gs} characteristics. And, the higher output power (higher $i_{D\rightarrow G}$) will enhance the difference.

With the help of the proposed models, v_{gs} characteristics of the upper-side and lower-side were simulated and compared with the experimental results. The pulling down amplitude of v_{gs} spike in the lower-side increases as the output power rises in both simulation and experiment, as well as the amplitude of the oscillation in the upper-side. Therefore, the trend of changes in the v_{gs} characteristics along with the increasing output power from simulation are consistent with that of the experimental results.

In addition, different conditions of gate resistance for the SiC power module are compared. A higher gate resistance can reduce dv_{DS}/dt, thus the v_{gs} spike issue and oscillation can be alleviated. Based on the proposed models, the trend of changes in the v_{gs} characteristics along with the increasing gate resistance can be simulated, and the results are shown to be consistent with that of the experimental results. This further confirms the validity of the proposed equivalent RLC circuit model and the rationality of the analysis about the mechanisms behind the v_{gs} characteristics at turn-on transient for the SiC half-bridge power module. Based on the model, a proper design of the gate driver circuit matched with the power module design and output power level can be obtained in a short design cycle.

Author Contributions: Conceptualization, M.Z.; formal analysis, N.R.; funding acquisition, Q.G. and K.S.; project administration, Q.G.; resources, X.Z.; supervision, K.S.; validation, K.S.; writing—original draft, M.Z.; writing—review and editing, M.Z., N.R., J.Z. and K.S. All authors have read and agreed to the published version of the manuscript.

Funding: This study is supported by the State Key Laboratory of Advanced Power Transmission Technology (Grant No. GEIRI-SKL-2018-009) and National Key Research & Development Program of China (2018YFB0905700).

Conflicts of Interest: The authors declare no conflict of interest.

References

1. Josifovic, I.; Popovic-Gerber, J.; Ferreira, J.A. Improving SiC JFET Switching Behavior Under Influence of Circuit Parasitics. *IEEE Trans. Power Electron.* **2012**, *27*, 3843–3854. [CrossRef]
2. Wrzecionko, B.; Bortis, D.; Biela, J.; Kolar, J.W. Novel AC-Coupled Gate Driver for Ultrafast Switching of Normally Off SiC JFETs. *IEEE Trans. Power Electron.* **2011**, *27*, 3452–3463. [CrossRef]
3. Lechler, M.; Piepenbreier, B. Regenerative SiC frequency converter with compact Z-source DC-link and sinusoidal output voltage. In Proceedings of the International Symposium on Power Electronics, Electrical Drives, Automation and Motion, Sorrento, Italy, 20–22 June 2012; pp. 1506–1511.
4. Anthony, P.; McNeill, N.; Holliday, D. High-speed resonant gate driver with controlled peak gate voltage for silicon carbide MOSFETs. In Proceedings of the 2012 IEEE Energy Conversion Congress and Exposition (ECCE), Raleigh, NC, USA, 15–20 September 2012; pp. 2961–2968.
5. Jauregui, D. *Reducing Ringing through PCB Layout Techniques*; Technical Report; Texas Instruments: Dallas, TX, USA, 2009.
6. Dong, Z.; Wu, X.; Sheng, K.; Zhang, J. Impact of common source inductance on switching loss of SiC MOSFET. In Proceedings of the 2015 IEEE 2nd International Future Energy Electronics Conference (IFEEC), Taipei, China, 1–4 November 2015; pp. 1–5.
7. Rondon-Pinilla, E.; Morel, F.; Vollaire, C.; Schanen, J.L. Modeling of a Buck Converter with a SiC JFET to Predict EMC Conducted Emissions. *IEEE Trans. Power Electron.* **2013**, *29*, 2246–2260. [CrossRef]
8. Czuchra, W.; Mysiński, W.; Woszczyna, B. Electromagnetic compatibility of SiC technology power converter. *Tech. Trans.* **2016**, *5*, 87–98.
9. Rondon, E.; Morel, F.; Vollaire, C.; Schanen, J.L. Impact of SiC components on the EMC behaviour of a power electronics converter. In Proceedings of the 2012 IEEE Energy Conversion Congress and Exposition (ECCE), Raleigh, NC, USA, 15–20 September 2012; pp. 4411–4417.
10. Robutel, R.; Martin, C.; Buttay, C.; Morel, H.; Mattavelli, P.; Boroyevich, D.; Meuret, R. Design and Implementation of Integrated Common Mode Capacitors for SiC-JFET Inverters. *IEEE Trans. Power Electron.* **2013**, *29*, 3625–3636. [CrossRef]
11. Yin, S.; Tseng, K.J.; Tong, C.F.; Simanjorang, R.; Gajanayake, C.J.; Gupta, A.K. A novel gate assisted circuit to reduce switching loss and eliminate shoot-through in SiC half bridge configuration. In Proceedings of the 2016 IEEE Applied Power Electronics Conference and Exposition (APEC), Long Beach, CA, USA, 20–24 March 2016; pp. 3058–3064.
12. Yin, S.; Tseng, K.J.; Tong, C.F.; Simanjorang, R.; Gajanayake, C.J.; Nawawi, A.; Liu, Y.; Liu, Y.; See, K.Y.; Sakanova, A.; et al. Gate driver optimization to mitigate shoot-through in high-speed switching SiC half bridge module. In Proceedings of the 2015 IEEE 11th International Conference on Power Electronics and Drive Systems, Sydney, Australia, 9–12 June 2015; pp. 484–491.
13. Zhang, Z.; Wang, F.; Tolbert, L.M.; Blalock, B.J. Active Gate Driver for Crosstalk Suppression of SiC Devices in a Phase-Leg Configuration. *IEEE Trans. Power Electron.* **2013**, *29*, 1986–1997. [CrossRef]
14. Zhang, W.; Zhang, L.; Mao, P.; Chan, X. Analysis of SiC MOSFET Switching Performance and Driving Circuit. In Proceedings of the 2018 IEEE International Power Electronics and Application Conference and Exposition (PEAC), Shenzhen, China, 4–7 November 2018; pp. 1–4.
15. Li, H.; Zhong, Y.; Yu, R.; Yao, R.; Long, H.; Wang, X.; Huang, Z. Assist Gate Driver Circuit on Crosstalk Suppression for SiC MOSFET Bridge Configuration. *IEEE J. Emerg. Sel. Top. Power Electron.* **2019**, 1. [CrossRef]
16. Zhang, W.; Zhang, Z.Y.; Wang, F.; Costinett, D.; Tolbert, L.; Blalock, R. Common source inductance introduced self-turn-on in MOSFET turn-off transient. In Proceedings of the 2017 IEEE Applied Power Electronics Conference and Exposition (APEC), Tampa, FL, USA, 26–30 March 2017; pp. 837–842.
17. Lemmon, A.; Mazzola, M.; Gafford, J. Instability in half-bridge circuits switched with wide band-gap transistors'. *IEEE. Trans. Power Electron.* **2014**, *28*, 2380–2392. [CrossRef]
18. Alatise, O.; Parker-Allotey, N.A.; Hamilton, D.; Mawby, P. The Impact of Parasitic Inductance on the Performance of Silicon–Carbide Schottky Barrier Diodes. *IEEE Trans. Power Electron.* **2012**, *27*, 3826–3833. [CrossRef]

19. Yang, F.; Wang, Z.J.; Zhang, Z.; Campbell, S.; Wang, F.; Chinthavali, M. Understanding middle-point inductance's effect on switching transients for multi-chip SiC package design with P-cell/N-cell concept. In Proceedings of the 2018 IEEE Applied Power Electronics Conference and Exposition (APEC), San Antonio, TX, USA, 4–8 March 2018; pp. 1742–1749.
20. Liu, T.; Wong, T.T.Y.; Shen, Z.J. A Survey on Switching Oscillations in Power Converters. *IEEE J. Emerg. Sel. Top. Power Electron.* **2019**, 1. [CrossRef]
21. Middelstaedt, L.; Lindemann, A. Optimization of critical oscillations within a boost converter based on an analytical model. In Proceedings of the 2016 18th European Conference on Power Electronics and Applications (EPE'16 ECCE Europe), Karlsruhe, Germany, 5–9 September 2016; pp. 1–9.
22. Jahdi, S.; Alatise, O.; Alexakis, P.; Ran, L.; Mawby, P. The Impact of Temperature and Switching Rate on the Dynamic Characteristics of Silicon Carbide Schottky Barrier Diodes and MOSFETs. *IEEE Trans. Ind. Electron.* **2014**, *62*, 163–171. [CrossRef]
23. Liu, T.; Ning, R.; Wong, T.T.Y.; Shen, Z.J. Modeling and Analysis of SiC MOSFET Switching Oscillations. *IEEE J. Emerg. Sel. Top. Power Electron.* **2016**, *4*, 747–756. [CrossRef]
24. Wang, J.; Chung, H.S.H.; Li, R.T.H. Characterization and Experimental Assessment of the Effects of Parasitic Elements on the MOSFET Switching Performance. *IEEE Trans. Power Electron.* **2012**, *28*, 573–590. [CrossRef]
25. Zhao, S.; Dearien, A.; Wu, Y.; Farnell, C.; Rashid, A.U.; Luo, F.; Mantooth, H.A. Adaptive Multi-Level Active Gate Drivers for SiC Power Devices. *IEEE Trans. Power Electron.* **2019**, *35*, 1882–1898. [CrossRef]
26. Lobsiger, Y.; Kolar, J.W. Closed-loop di/dt and dv/dt IGBT gate driver. *IEEE Trans. Power Electron.* **2015**, *30*, 3402–3417. [CrossRef]
27. Zhang, B.F.; Xu, J.M.; Qian, Q.; Zhang, Z.; Xie, S.J. Analysis on Characteristics of SiC-MOSFET and Key Techniques of Its Applications. *J. Power Supply* **2016**, *14*, 39–51.
28. Zhou, Q.; Gao, F. A gate driver of SiC MOSFET for suppressing the negative voltage spikes in a bridge circuit. *IEEE Trans. Power Electron.* **2018**, *33*, 2339–2353.
29. Zeng, Z.; Li, X. Comparative Study on Multiple Degrees of Freedom of Gate Drivers for Transient Behavior Regulation of SiC MOSFET. *IEEE Trans. Power Electron.* **2018**, *33*, 8754–8763. [CrossRef]
30. Chen, Z.; Boroyevich, D.; Burgos, R. Experimental parametric study of the parasitic inductance influence on MOSFET switching characteristics. In Proceedings of the 2010 International Power Electronics Conference—ECCE ASIA-, Sapporo, Japan, 21–24 June 2010; pp. 164–169.
31. Wang, Z.H.; Zhang, J.M.; Wu, X.K. Analysis of stray inductance's influence on SiC MOSFET switching performance. In Proceedings of the Energy Conversion Congress and Exposition (ECCE), Pittsburgh, PA, USA, 14–18 September 2014; pp. 2838–2843.
32. Ke, J.J.; Zhao, Z.B.; Wei, C.J.; Xu, P.; Xie, Z.K.; Yang, F. Effect of the Parasitic Inductance on SiC MOSFET Switching Characteristics. *Semicond. Technol.* **2017**, *42*, 194–199.
33. Yang, B.; Zhang, J. Effect and utilization of common source inductance in synchronous rectification. In Proceedings of the Twentieth Annual IEEE Applied Power Electronics Conference and Exposition, APEC 2005, Austin, TX, USA, 6–10 March 2005; Volume 3, pp. 1407–1411.

micromachines

Article

Silicon Carbide Microstrip Radiation Detectors

Donatella Puglisi [1,2,*] **and Giuseppe Bertuccio** [1,3]

[1] Department of Electronics, Information and Bioengineering, Politecnico di Milano, Campus Como, 22100 Como, Italy; giuseppe.bertuccio@polimi.it
[2] Department of Physics, Chemistry and Biology, Sensor and Actuator Systems, Linköping University, 58183 Linköping, Sweden
[3] Italian National Institute of Nuclear Physics (INFN), Section Milano, 20133 Milan, Italy
* Correspondence: donatella.puglisi@liu.se; Tel.: +46-13-28-1273

Received: 17 July 2019; Accepted: 27 November 2019; Published: 30 November 2019

Abstract: Compared with the most commonly used silicon and germanium, which need to work at cryogenic or low temperatures to decrease their noise levels, wide-bandgap compound semiconductors such as silicon carbide allow the operation of radiation detectors at room temperature, with high performance, and without the use of any bulky and expensive cooling equipment. In this work, we investigated the electrical and spectroscopic performance of an innovative position-sensitive semiconductor radiation detector in epitaxial 4H-SiC. The full depletion of the epitaxial layer (124 μm, 5.2×10^{13} cm^{-3}) was reached by biasing the detector up to 600 V. For comparison, two different microstrip detectors were fully characterized from −20 °C to +107 °C. The obtained results show that our prototype detector is suitable for high resolution X-ray spectroscopy with imaging capability in a wide range of operating temperatures.

Keywords: silicon carbide; semiconductor radiation detector; microstrip detector

1. Introduction

The concept of using compound semiconductors as radiation detectors was introduced in 1945 by Van Heerden [1,2], who was the first to be able to detect alpha and gamma rays with solid-state radiation counters. His pioneering results gave rise to a new class of radiation detectors, which is now commonly known as semiconductor detectors. Compared to gas detectors, semiconductor detectors require much lower average energies for the creation of electron-hole pairs (30 eV for gas [3], 3.7 eV for Si [4], 7.8 eV for 4H-SiC [5]), which bring higher energy resolution in radiation spectroscopy [6,7]. Since the 1960s, the most commonly used semiconductor materials have been high-purity silicon (Si) and germanium (Ge), the main limitation of which is that they must operate at liquid nitrogen temperature. Since the 1990s, intense research activity has been carried out on other semiconductors for manufacturing detectors able to operate at room temperature, such as gallium arsenide (GaAs), cadmium telluride (CdTe), and cadmium zinc telluride (CdZnTe) [8–12]. In the last two decades, silicon carbide (SiC) has obtained increasing interest in the field of radiation detectors due to the achievement of a high purity level in the crystal structure and considerable thickness (>100 μm) in the epitaxial layer. This finally achieved recognition for semiconductor detectors as a real alternative to Si-based radiation detectors, which present possibilities but also limitations at and above room temperature, as well as in high-radiation environments [13,14]. There are certain properties that make SiC especially suitable for the realization of ionizing radiation detectors. Thanks to the wide energy bandgap of the polytype 4H-SiC (3.26 eV), which is three times higher than that of Si (1.12 eV), electronic devices fabricated in such material can operate at extremely high temperatures without suffering from negative effects, due to thermally generated charge carriers [15]. Silicon carbide radiation detectors benefit from this property because the wide energy bandgap allows the achievement of very low leakage currents, i.e., very low noise levels, even at the high electric fields applied during their operation. Moreover, the high thermal conductivity of 4H-SiC (3.8 W/cm°C) enables SiC devices to dissipate large amounts of excess generated heat, which would cause a temperature increase, responsible for degradation of the device's performance. High thermal conductivity is useful for increasing the radiation hardness of the detector, as well as for controlling the operating temperature when the front-end electronics are close to, or in contact with, the detector [16]. Furthermore, SiC can withstand an internal electric field over eight to ten times greater than GaAs or Si (2 MV/cm for 4H-SiC vs. 0.4 MV/cm for GaAs or 0.3 MV/cm for Si) without undergoing avalanche breakdown. This property enables the fabrication of very high-voltage devices [17]. In the case of X-ray detection and spectroscopy, the high breakdown field of 4H-SiC allows, in principle, the detector to work always in the regime of saturated-electron and hole-drift velocities, independently of the detector's active region width. When this operation condition is coupled with epitaxial material of high crystalline quality, a full and fast charge collection can be expected [16], as well as a high sensitivity, as already demonstrated [18]. Such properties allow SiC-based devices to be operated without any costly, bulky, and power-consuming cooling systems, as in the case of Si- or Ge-based devices, while maintaining an excellent signal-to-noise ratio over a wide range of temperatures. This leads to notable advantages in terms of the lower cost, more compact size, lighter weight, lower power consumption, and higher performance of SiC detectors. Further explanation of the electrical properties of SiC in connection with the ionizing detector performance benefits can be found in [16].

Microstrip detectors find application where the position of the radiation interaction is necessary information for the physical process to be studied. The advantage of using microstrips with respect to other position-sensitive detectors, such as pixel detectors, is a lower number of readout channels. Several microstrip detectors have been developed in Si for high-energy physics, or in Ge, CdTe and GaAs for X-ray spectroscopy [19–22]. In this work, we investigated the electrical and spectroscopic performance of two innovative position-sensitive radiation detectors in epitaxial 4H-SiC, using microstrip geometry. The detectors were characterized in detail at different temperatures and applied bias voltages. The obtained results are presented and discussed in the following sections.

2. Materials and Methods

Two different designs of silicon carbide microstrip detectors have been realized on top of two-inch high-purity epitaxial 4H-SiC wafer produced by LPE Epitaxial Technology Center [23]. Each detector consists of 32 strips with a length of 2 mm, a width of either 25 μm (SM1) or 50 μm (SM3), and a pitch of either 55 μm (SM1) or 100 μm (SM3)—see Figure 1. Each of these strips can be read out independently by a front-end electronics channel, and therefore behaves as a separate detector. A cross-sectional view of the 4H-SiC microstrip structure is shown in Figure 2. The SiC epitaxial layer, which is the active region of the detector, has a maximum thickness of 124 μm, as experimentally measured (Figure 3).

The application of a reverse bias at the common back electrode (ohmic contact) creates a depletion region, x_d, depending on the applied bias, V_R, and the residual doping (donor) concentration, N_D, within the material, according to

$$x_d = \sqrt{\frac{2\varepsilon_0\varepsilon_r}{qN_D}\left(\psi_{bi} - V_R - \frac{kT}{q}\right)} \quad (1)$$

where ψ_{bi} is the built-in potential and the term $\frac{kT}{q}$ arises from the contribution of the majority-carrier distribution tail [24].

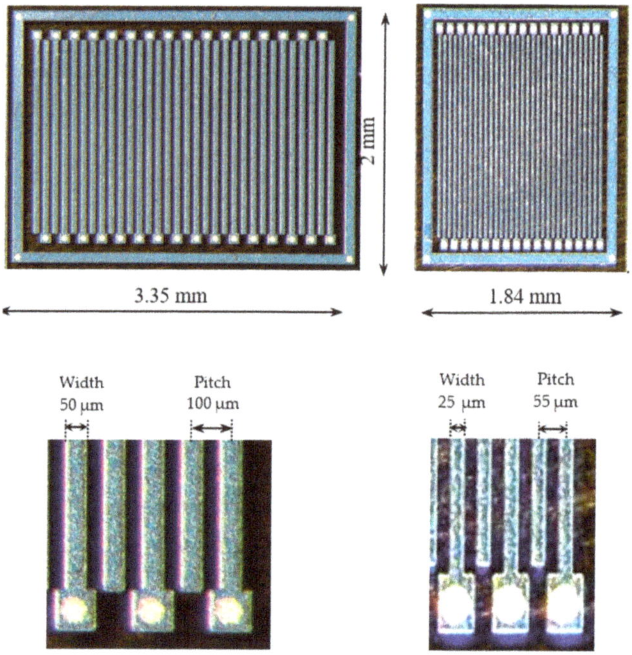

Figure 1. Photographs of the two microstrip detectors used in this work, together with a detail of their peripheral regions with bonding pads.

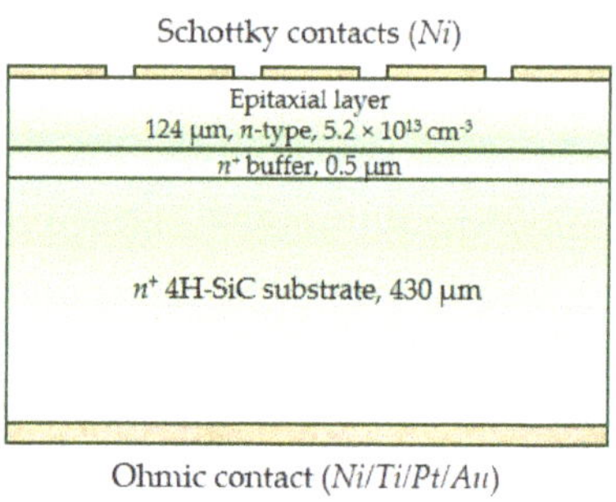

Figure 2. Cross-sectional view of the 4H-SiC microstrip structure.

The residual doping concentration depends on the homogeneity of the epitaxial layer. The donor concentration profile, $N_D(x)$, was determined as a function of the depleted layer width from capacitance-voltage measurements, as described in Section 3.

3. Results

3.1. Electrical Characterization

3.1.1. Capacitance–Voltage Characterization

Capacitance–voltage (C–V) measurements were carried out up to 600 V at 25 °C, in order to determine the donor-concentration profile of the epitaxial layer (Figure 3a). The detector was placed in a test fixture Agilent 16065A connected to an Agilent 4284A Precision LCR Meter (Santa Clara, CA, USA). A Keithley 2410 voltage source (Cleveland, OH, USA), operating in the four-wire connection mode, was used to bias the device and measure the applied voltage. The measurement was performed with a 100 mV AC signal at 100 kHz. The donor-concentration profile as a function of the depleted layer width was determined from the slope of a $1/C^2$–V curve, according to [24], see Figure 3a. Please note that the C–V measurements were performed using a 4H-SiC Schottky diode with area A = 5 mm^2, produced from the same wafer. A full depletion of 124 μm was reached, polarizing the detector up to 600 V (Figure 3b). A mean value of $\langle N_D \rangle = (5.20 \pm 0.06) \times 10^{13}$ cm^{-3} was determined (Figure 4).

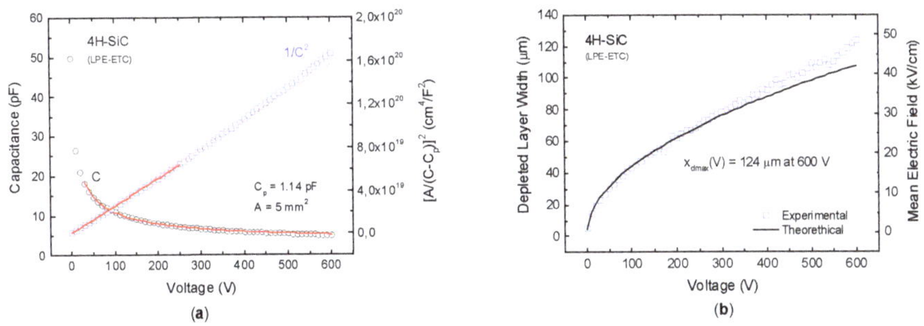

Figure 3. (**a**) Capacitance–voltage (C–V) and $1/C^2$–V per unit area characteristics. From the fit curve, the mean value of the donor concentration is expected to be $(5.56 \pm 0.05) \times 10^{13}$ cm^{-3}; (**b**) depleted layer and mean electric field as a function of applied voltage, as derived from C–V measurements. The theoretical result is obtained using 5.56×10^{13} cm^{-3} as the mean value of the doping concentration.

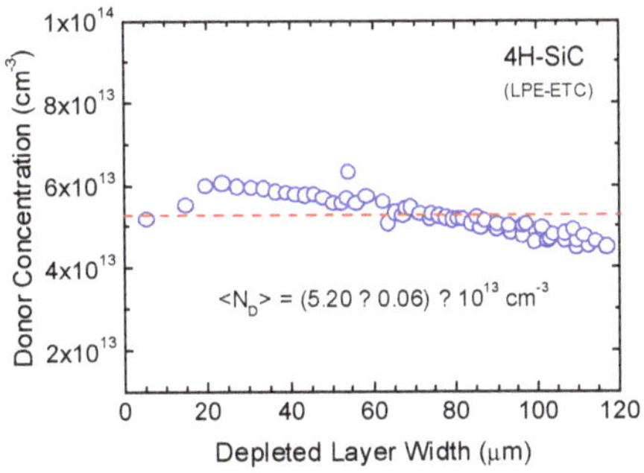

Figure 4. Donor concentration profile as a function of the depleted layer width. The full depletion of 124 μm was reached at 600 V. A mean value of $\langle N_D \rangle = 5.2 \times 10^{13}$ cm^{-3} was determined.

3.1.2. Statistical Leakage–Current Distribution

Current–voltage (I–V) measurements were carried out on each of the 32 strips of two different SiC detectors, SM1 and SM3, at room temperature (Figure 5). The two detectors were biased at 100 V and 200 V from the back ohmic contact using a Keithley 2410 source meter, whereas the current of each strip was measured connecting a Keithley 6430 electrometer to the front, rectifying the Schottky contact. The guard electrode surrounding the microstrips was kept to ground to collect the parasitic current generated at the device's chip edges. Figure 5 shows the current and current density values for each of the 32 strips of the two different microstrip detectors as measured at 25 °C and 200 V. Such bias voltage generates an inner electric field of about 30 kV/cm. Ultra-low current mean values of 2.2 fA and 7.6 fA were measured for the two microstrips under test, corresponding to current densities of 4.4 pA/cm^2 and 15.2 pA/cm^2.

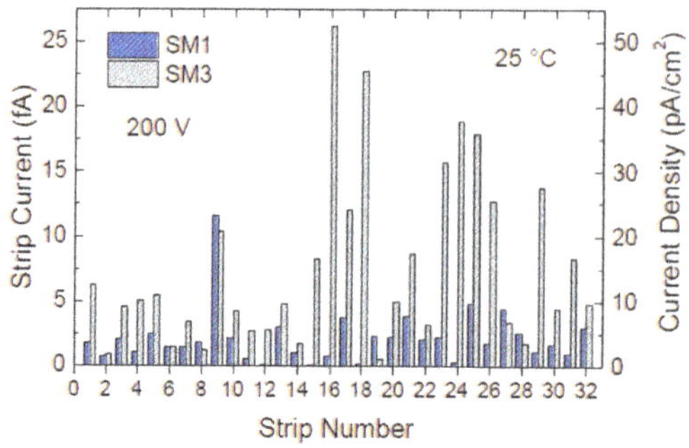

Figure 5. Current and current density measured at 25 °C and 200 V on the 32 strips of the two microstrip detectors, SM1 and SM3. Current values of few fA (current densities of low pA/cm^2) were measured on all strips.

3.1.3. Temperature Dependence

The temperature dependence of the strip current as a function of reverse-bias voltage is shown in Figure 6. From now on, only results obtained with the detector SM1 are presented. In order to perform this measurement, the device was attached to a Teflon circuit board using a silver conductive glue. Electrical contacts were established using 25 µm gold wire-bonding connections. Measurements were acquired, biasing up to 200 V the back contact with the Keithley 2410 source meter and reading the currents from the Keithley 6430 electrometer. Tests were carried out inside an environmental chamber, setting the temperature at 27 °C, 47 °C, 67 °C, 87 °C, and 107 °C, and monitoring it by means of a thermocouple placed near to the device. During each measurement, the temperature changes were monitored within ±0.1 °C. The current density was calculated considering a strip active area of 5×10^{-4} cm^{-2}.

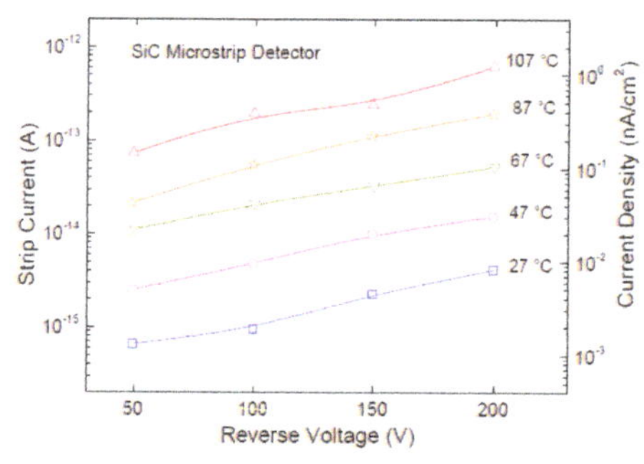

Figure 6. Current and current density dependence from temperature in the range 27 to 107 °C.

Such a thermally activated process is described by the Arrhenius plot that, according to the emission theory, is expressed by

$$I = I_0 \cdot \exp\left[\left(-\frac{E_A}{kT}\right)\cdot\left(1 - \frac{T}{T_0}\right)\right] \quad (2)$$

where I_0 is the saturation current, T_0 is the room temperature, and E_A is the activation energy [5,25].

Figure 7 shows the Arrhenius plots of the leakage current as a function of 1000/T at four different reverse voltages, i.e., 50 V, 100 V, 150 V, and 200 V. The activation energy is given by the slope of linear fit of data. Values from 0.57 eV to 0.65 eV were calculated in the voltage range 50 V to 200 V. According to the literature, these values refer to major deep levels ($Z_{1/2}$ center) within the bandgap [26–28].

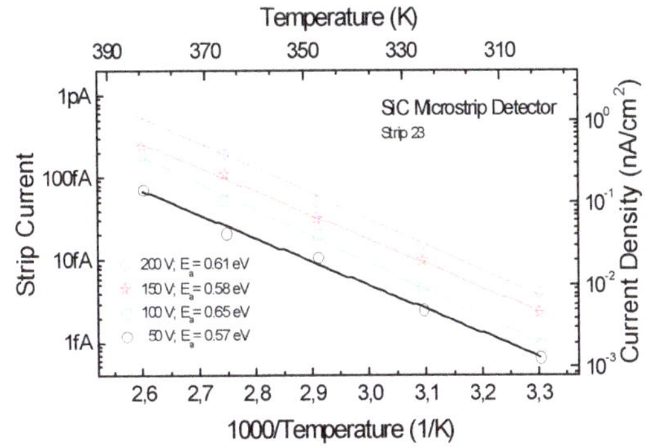

Figure 7. Arrhenius plots of the leakage current (and current density) versus the reverse of temperature at four different voltages. The activation energy values are from 0.57 eV to 0.65 eV in the voltage range 50 to 200 V.

3.1.4. Interstrip Resistance Measurements

Current–voltage measurements shown in previous sections refer to the characterization of single strips, considering that the measured current arrived only from the back contact of the device, and possible latent currents from adjacent strips were negligible. In order to determine the bias limit condition so that two adjacent strips can be considered isolated, we measured the interstrip resistance.

Interstrip resistance measurements were carried out by measuring the current between two consecutive strips, keeping the back contact at 100 V and the guard at 0 V. One of the two strips (microstrip 2 in Figure 8) is biased from −5 V to +5 V while the current of the other strip (microstrip 1 in Figure 8), kept at 0 V, is measured by an electrometer. We repeated the test on three couples of strips. The negative slope shown in Figure 8 is due to the application of the bias voltage to microstrip 2 while measuring the current at microstrip 1. The mean value of resistance between two adjacent strips of SM1 resulted in 5.3 TΩ.

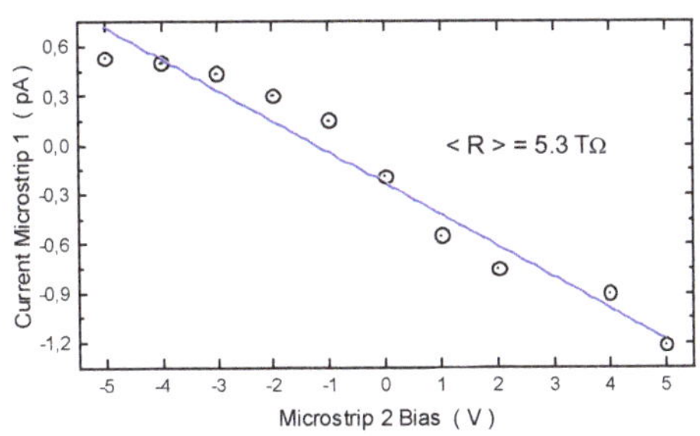

Figure 8. Mean value of resistance between two adjacent strips.

3.2. X-Ray Spectroscopy

3.2.1. Room Temperature

The detector was irradiated with a 397 kBq ^{241}Am source placed at a few cm from the detector surface. Figure 9a shows a ^{241}Am spectrum acquired using a SiC microstrip detector at 21 °C. The spectrum was acquired at 200 V reverse-bias condition using 12.8 µs peaking time in the triangular signal processing. The pulser line width is 214 eV full width at half maximum (FWHM) corresponding to an equivalent noise charge of 11.6 electrons root mean square (rms). Figure 9b shows a detail of the same measurement up to 28 keV. Several X-ray lines from Mn, Cu, Np, and Ag can be clearly distinguished with a very good resolution, i.e., enough to separate the *K* and *L* lines of neighboring elements. Conventionally, the energy resolution, that is the FWHM, is specified for the Mn $K\alpha$ peak at 5.9 keV, which is 213 eV for our SiC microstrip detector at room temperature (Figure 9). It is notable that Si(Li) and silicon drift detectors can achieve 130–150 eV FWHM, and Ge detectors can even achieve 115 eV FWHM for the Mn $K\alpha$ peak at 5.9 keV, but with liquid-nitrogen cooling [29].

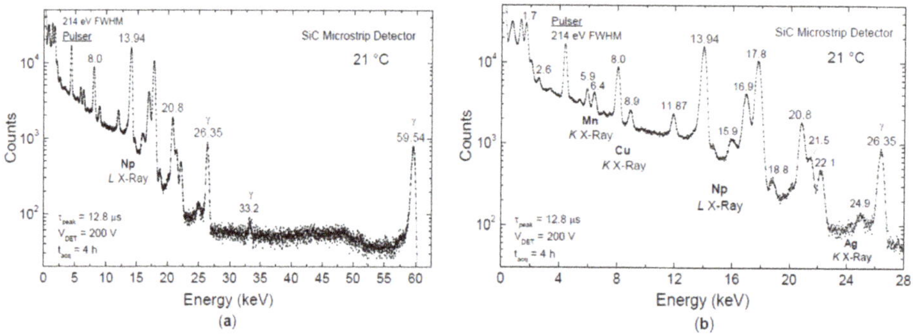

Figure 9. (**a**) X-ray spectrum from a ^{241}Am source acquired at 21 °C using the SiC microstrip detector SM1 and an ultra-low noise front-end (PRE5 no. 3); (**b**) detailed X-ray spectroscopy in the energy range 0 to 28 keV.

The analysis of linearity calculated on seven well-resolved peak lines at 8.0, 11.87, 13.94, 17.8, 20.8, 26.35 and 59.54 keV is shown in Figure 10. The percentage error from linearity is below ±0.05%.

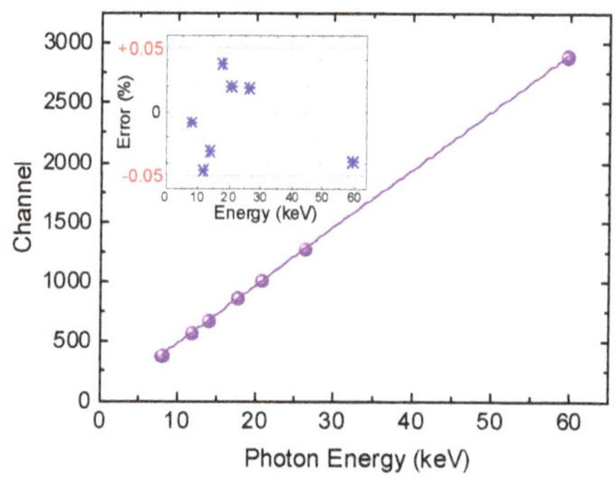

Figure 10. Analysis of linearity calculated on seven well-resolved peak lines, as shown in Figure 9. The percentage error from linearity is below ±0.05%.

3.2.2. Dependence of X-Ray Response on Detector Bias

The dependence of X-ray response on detector voltage was explored by biasing the detector from 10 V to 200 V at 25 °C and using a peaking time of 12.8 μs. A comparison between two spectra acquired at 10 V (bottom, blue) and 200 V (top, red) is shown in Figure 11. The pulser centroid is stable at both 10 V and 200 V. The peak at 13.94 keV shows a small shift of six channels (from 669 at 10 V to 675 at 200 V) which corresponds to 124 eV (Figure 12). The two peaks, as compared in Figure 12a, show a Gaussian symmetry without any tails. This experiment shows that SiC detectors can be operated in a wide range of bias voltages without suffering from a strong performance loss. Figure 12b shows the detection rate of the 13.94 keV photon peak at six different applied reverse-bias voltages, V_b. As expected, the photon rate increases with the square root of V_b due to the widening of the active region (depletion layer).

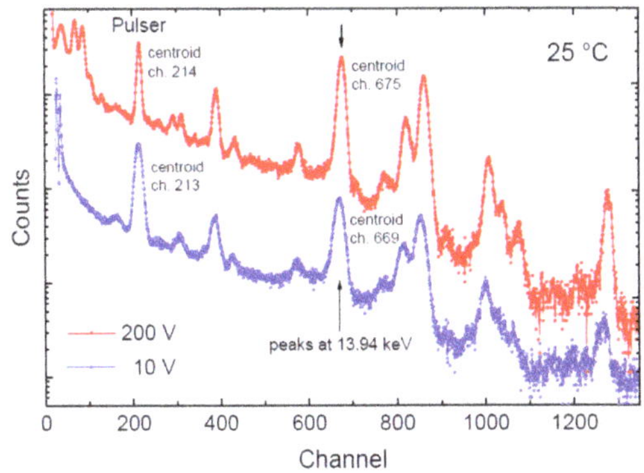

Figure 11. X-ray spectra from a ^{241}Am source acquired at 25 °C and at 10 V (bottom) and 200 V (top).

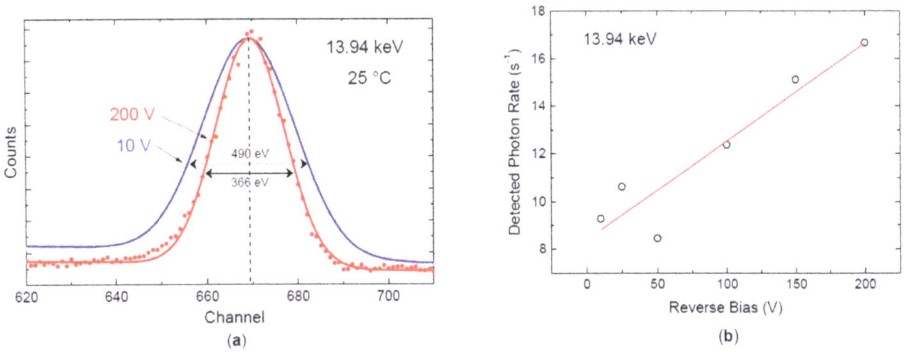

Figure 12. (**a**) Comparison between the two peaks at 13.94 eV obtained at 10 V and 200 V of applied voltage. The Gaussian symmetry without tails can be noticed; (**b**) Detected 13.94 keV photon rate as a function of the applied reverse bias.

3.2.3. High Statistics and Temperature Dependence

Figure 13 shows the results obtained by acquiring an X-γ-ray spectrum for 10 h at 80 V reverse-bias condition, and maintaining the thermostatic chamber at a constant temperature of 30 °C. Figure 13a compares spectra acquired after almost 2 h and after 10 h. The very small broadening of the pulser and emission lines should be noted, which demonstrates very good stability of the detector response to X-ray exposure. The analysis of linearity on *K* Cu and *L* Np X-ray monoenergetic lines shows a very small linearity error within ±0.04% after 10 h of acquisition (Figure 13b).

Figure 14 shows a comparison between X-ray spectra acquired at three different temperatures, i.e., −20 °C, +30 °C, and +80 °C. As expected, the width of pulser and emission lines increased by increasing the operating temperature: the FWHM of the pulser changed from 205 eV at −20 °C, to 215 eV at +30 °C, and 249 eV at +80 °C. It is worth noticing the small broadening of the lines at +80 °C, which demonstrates the suitability of our microstrip detector to be used at high temperatures with very good stability of the detector response.

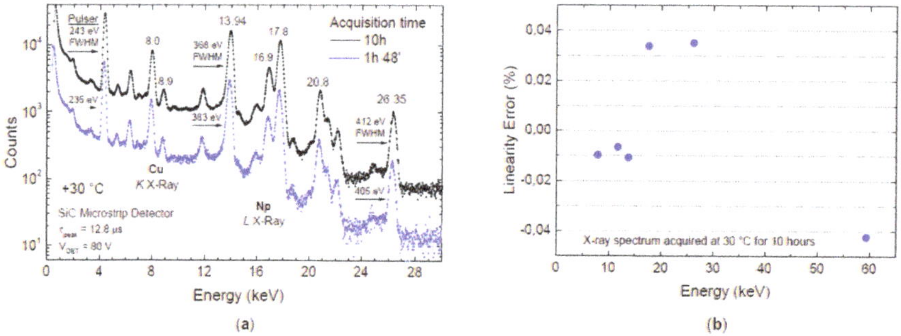

Figure 13. (**a**) X-ray spectra from ^{241}Am source after almost 2 and 10 h of acquisition at 30 °C in a thermostatic chamber. The very small broadening of pulser and emission lines demonstrates very good stability of the detector response to X-ray exposure; (**b**) linearity error after 10 h of acquisition based on *K* Cu and *L* Np X-ray monoenergetic lines.

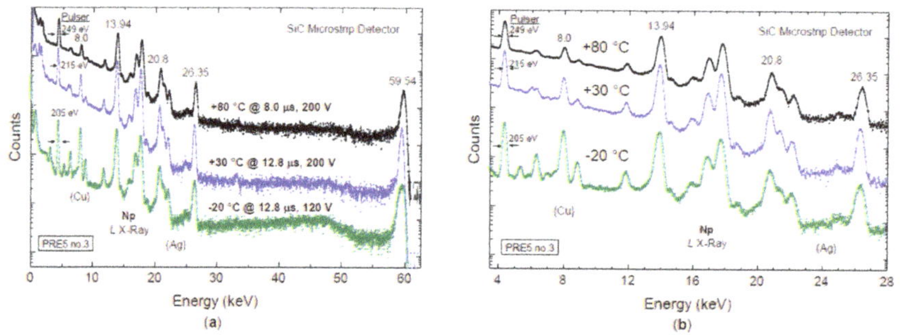

Figure 14. (**a**) Comparison between three X-ray spectra acquired at −20 °C, +30 °C, and +80 °C in the range 0–60 keV; (**b**) Detail of the X-ray spectra in the range 4–28 keV. Note a small broadening of the pulser line by increasing the operating temperature.

4. Discussion

Two SiC-based microstrip detectors were fully characterized in a wide temperature range by means of electrical and spectroscopic measurements. A high stability of the detector response as a function of operating temperature as well as of applied voltage was widely demonstrated.

The very low leakage currents (current densities) from about 2 fA (4 pA/cm^2) at 25 °C to 620 fA (1.2 nA/cm^2) at 107 °C are among the best values measured on SiC detectors, and more than one order of magnitude lower than most silicon detectors [30,31]. Since the shot noise of the leakage current is a significant noise contribution in a radiation spectroscopy system, we can say that our SiC detectors allow the achievement of high signal-to-noise ratios. A good isolation between adjacent strips was demonstrated by the high value of the measured interstrip resistance of 5.3 TΩ, confirming that possible latent currents from adjacent strips can be considered negligible.

A very good doping uniformity of the whole epitaxial layer was also demonstrated. For the first time, a full depletion of 124 µm was reached, polarizing the detector at 600 V, and a mean value of $\langle N_D \rangle = 5.2 \times 10^{13}$ cm^{-3} was determined. In comparison with previous studies, this result is the best observed [32].

The X-ray spectra acquired from a ^{241}Am source at different voltages, temperatures, and exposure times showed high stability and a high spectroscopic resolution under all tested experimental conditions.

Different voltages were used to verify the effect of the applied bias voltage on the device performance. The spectroscopic response of our SiC detector does not significantly depend on the bias voltage, as shown in Figure 11, where two extreme bias voltages (10 V and 200 V) were used.

No tails in the spectral lines were observed, which means that no significant charge trapping occurred in these devices. Remarkably, no strong performance loss was observed at 10 V, and substantially no difference was observed under operation between 80 V and 200 V. The possibility of using a lower voltage without losing significant information is an advantage for those applications wherein lower power consumption is desirable. The better resolution obtained at 200 V is due to a lower capacitance. We avoided operating the SiC detector above 200 V to prevent the risk of possible damages due to accidental breakdown or electrostatic discharge. Also, the exposure time to the ^{241}Am source does not affect the spectroscopic performance of our SiC detector, as demonstrated by negligible differences in the peak resolution after almost 2 and 10 h of acquisition (Figure 13). This means that it is not necessary to wait for a long time before getting all the main information from the device.

Finally, it is worth noticing the high resolution and very good stability in the performance of our SiC microstrip detector between −20 °C and +80 °C (Figure 14), which pave the way for use in a wide range of applications that are prohibitive for other conventional semiconductor detectors.

5. Conclusions

Even considering an initial higher cost for the SiC material in comparison to Si or Ge, SiC-based semiconductor radiation detectors are advantageous for use in operating conditions under which conventional semiconductor detectors in Si or Ge cannot adequately perform. One overall advantage is the elimination of cryogenic or Peltier cooling systems that allows the fabrication of far more compact, more stable, lighter, and lower power radiation detector systems. This also implies, as a direct consequence, economic advantages that go beyond the mere costs for the material itself.

Our findings confirm the high quality and the good uniformity of the epitaxial layer used for manufacturing our SiC-based prototypes, as well as the suitability of such devices to be used as high resolution X-ray detectors over a wide range of operating temperatures.

Author Contributions: Investigation, D.P. and G.B.; Supervision, G.B.; Writing – original draft, D.P. and G.B.; Writing – review & editing, D.P.

Funding: This research was funded by the Italian National Institute of Nuclear Physics (INFN) and by the Italian Space Agency (ASI). The APC was funded by Linköping University.

Acknowledgments: All experiments were performed at the Semiconductor and Integrated Circuit Laboratory of the Department of Electronics, Information and Bioengineering, Politecnico di Milano, Como Campus, Italy. The authors would like to thank C. Lanzieri for device manufacturing process at Selex Integrated Systems (Rome), P. Ferrari for his contribution to the electrical characterization, S. Masci for the device bonding, S. Caccia and D. Macera for their collaboration for the readout electronics. The authors would like to dedicate this work to the everlasting memory of Andrea Puglisi.

Conflicts of Interest: The authors declare no conflicts of interest. The founding sponsors had no role in the design of the study; in the collection, analyses, or interpretation of data; in the writing of the manuscript, and in the decision to publish the results.

References

1. Van Heerden, P.J. The Crystal Counter: A new apparatus in nuclear physics for the investigation of β- and γ-rays. Part, I. *Physica* **1950**, *16*, 505–516. [CrossRef]
2. Van Heerden, P.J.; Milatz, J.M.W. The Crystal Counter: A new apparatus in nuclear physics for the investigation of β- and γ-rays. Part II. *Physica* **1950**, *16*, 517–527. [CrossRef]
3. Lerov, C.; Rancoita, P.-G. *Principles of Radiation Interaction in Matter and Detection*, 2nd ed.; Word Scientific: Singapore, 2009; p. 83. ISBN 978-981-4470-89-6.
4. Scholze, F.; Rabus, H.; Ulm, G. Mean energy required to produce an electron-hole pair in silicon for photons of energies between 50 and 1500 eV. *J. Appl. Phys.* **1998**, *84*, 2926–2939. [CrossRef]
5. Bertuccio, G.; Casiraghi, R. Study of silicon carbide for X-ray detection and spectroscopy. *IEEE Trans. Nucl. Sci.* **2003**, *50*, 175–185. [CrossRef]

6. Knoll, G.N. *Radiation Detection and Measurement*, 4th ed.; Wiley: New York, NY, USA, 2010; ISBN 978-0-470-13148-0.
7. Lutz, G. *Semiconductor Radiation Detectors: Device Physics*, 2nd printing of the 1st ed.; Springer: Berlin/Heidelberg, Germany, 2007; ISBN 978-3-540-71679-2.
8. McGregor, D.S.; Hermon, H. Room-temperature compound semiconductor radiation detectors. *Nucl. Instrum. Methods Phys. Res. A Accel. Spectrom. Detect. Assoc. Equip.* **1997**, *395*, 101–124. [CrossRef]
9. Bertuccio, G. Prospect for Energy Resolving X-ray Imaging with Compound Semiconductor Pixel Detectors. *Nucl. Instrum. Methods Phys. Res. A Accel. Spectrom. Detect. Assoc. Equip.* **2005**, *546*, 232–241. [CrossRef]
10. Schlesinger, T.E.; James, R.B. (Eds.) *Semiconductor for Room Temperature Nuclear Detectors Applications*, Semiconductor and Semimetal Series; 1st ed.; Academic Press: San Diego, CA, USA, 1995; Volume 43, ISBN 9780080571997.
11. Owens, A.; Bavdaz, M.; Peacock, A.; Poelaert, A. High resolution x-ray spectroscopy using GaAs arrays. *J. Appl. Phys.* **2001**, *90*, 5376–5381. [CrossRef]
12. Abbene, L.; Del Sordo, S.; Caroli, E.; Gerardi, G.; Raso, G.; Caccia, S.; Bertuccio, G. Hard X-ray Response of Pixellated CdZnTe detectors. *J. Appl. Phys.* **2009**, *105*, 124508. [CrossRef]
13. Bertuccio, G.; Pullia, A. Room Temperature X-Ray Spectroscopy with a Silicon Diode Detector and an Ultra Low Noise Preamplifier. *IEEE Trans. Nucl. Sci.* **1994**, *41*, 1704–1709. [CrossRef]
14. Pullia, A.; Bertuccio, G. Resolution Limits of Silicon Detectors and Electronics for Soft X Ray Spectroscopy at non Cryogenic Temperatures. *Nucl. Instrum. Methods Phys. Res. A Accel. Spectrom. Detect. Assoc. Equip.* **1996**, *380*, 1–5. [CrossRef]
15. Bertuccio, G.; Caccia, S.; Casiraghi, R.; Lanzieri, C. Possibility of sub-electron noise with room temperature silicon carbide pixel detectors. *IEEE Trans. Nucl. Sci.* **2006**, *53*, 2421–2427. [CrossRef]
16. Nava, F.; Bertuccio, G.; Cavallini, A.; Vittone, E. Silicon carbide and its use as a radiation detector material. *Meas. Sci. Technol.* **2008**, *19*, 102001. [CrossRef]
17. Kimoto, T.; Cooper, J.A. *Fundamentals of Silicon Carbide Technology*, 1st ed.; IEEE Press: Piscataway, NJ, USA; John Wiley & Sons: Hoboken, NJ, USA; Singapore Pte. Ltd.: Singapore, 2014; ISBN 978-1-118-31352-7.
18. Mandal, K.C.; Muzykov, P.G.; Terry, J.R. Highly sensitive x-ray detectors in the low-energy range on n-type 4H-SiC epitaxial layers. *Appl. Phys. Lett.* **2012**, *101*, 051111. [CrossRef]
19. Sadrozinski, H.F.-W. Silicon microstrip detectors in high luminosity application. *IEEE Trans. Nucl. Sci.* **1998**, *45*, 295–302. [CrossRef]
20. Rossi, G.; Morse, J.; Protic, D. Energy and position resolution of germanium microstrip detectors at X-ray energies from 15 to 100 keV. *IEEE Trans. Nucl. Sci.* **1999**, *46*, 765–773. [CrossRef]
21. Dusi, W.; Auricchio, N.; Brigliadori, L.; Cola, A.; Donati, A.; Landini, G.; Mengoni, D.; Perillo, E.; Siffert, P.; Ventura, G. A study of the spectroscopic performance of a CdTe microstrip detector. *IEEE Trans. Nucl. Sci.* **2003**, *50*, 1026–1030. [CrossRef]
22. Kaluza, A.; Ohms, T.; Rente, C.; Engels, R.; Reinartz, R.; Luth, H. Microstrip X-ray detector with a very high dynamic range based on LPE-GaAs. *IEEE Trans. Nucl. Sci.* **1998**, *45*, 724–727. [CrossRef]
23. LPE Epitaxial Technology, S.p.A. Available online: http://www.lpe-epi.com/ (accessed on 21 June 2019).
24. Sze, S.M.; Ng, K.K. *Physics of Semiconductor Devices*, 3rd ed.; John Wiley & Sons Inc.: Hoboken, NJ, USA, 2007; pp. 85–137. ISBN 978-0-471-14323-9.
25. Schroder, D.K. *Semiconductor Material and Device Characterization*, 3rd ed.; John Wiley & Sons Inc.: Hoboken, NJ, USA, 2006; p. 691. ISBN 978-0-471-73906-7.
26. Knoll, L.; Kranz, L.; Alfieri, G. Electrically active deep levels formed by thermal oxidation of n-type 4H-SiC. *J. App. Phys.* **2019**, *125*, 205302. [CrossRef]
27. Mannan, M.A.; Nguyen, K.V.; Pak, R.O.; Oner, C.; Mandal, K.C. Deep Levels in n-Type 4H-Silicon Carbide Epitaxial Layers Investigated by Deep-Level Transient Spectroscopy and Isochronal Annealing Studies. *IEEE Trans. Nucl. Sci.* **2016**, *63*, 1083–1090. [CrossRef]
28. Nava, F.; Castaldini, A.; Cavallini, A.; Errani, P.; Cindro, V. Radiation Detection Properties of 4H-SiC Schottky Diodes Irradiated up to 10^{16} n/cm^2 by 1 MeV Neutrons. *IEEE Trans. Nucl. Sci.* **2006**, *53*, 2977–2982. [CrossRef]
29. Statham, P.J. Recent Developments in Instrumentation for X-ray Microanalysis. In *Modern Developments and Applications in Microbeam Analysis*; Love, G., Nicholson, W.A.P., Armigliato, A., Eds.; Springer: Wien, Austria; New York, NY, USA, 1998; pp. 1–9. (Suppl. 15), ISBN 978-3-211-83106-9.

30. Ruddy, F.H.; Seidel, J.G.; Chen, H.; Dulloo, A.R.; Ryu, S.-H. High-Resolution Alpha-Particle Spectrometry Using 4H Silicon Carbide Semiconductor Detectors. *IEEE Trans. Nucl. Sci.* **2006**, *53*, 1713–1718. [CrossRef]
31. Bertuccio, G.; Caccia, S.; Nava, F.; Foti, G.; Puglisi, D.; Lanzieri, C.; Lavanga, S.; Abbondanza, G.; Crippa, D.; Preti, F. Ultra low noise epitaxial 4H-SiC X-ray detectors. *Mat. Sci. Forum* **2009**, *615*, 845–848. [CrossRef]
32. Bertuccio, G.; Puglisi, D.; Macera, D.; Di Liberto, R.; Lamborizio, M.; Mantovani, L. Silicon Carbide Detectors for in vivo Dosimetry. *IEEE Trans. Nucl. Sci.* **2014**, *61*, 961–965. [CrossRef]

micromachines

MDPI

Article

Investigation of the Young's Modulus and the Residual Stress of 4H-SiC Circular Membranes on 4H-SiC Substrates

Jaweb Ben Messaoud [1], Jean-François Michaud [1], Dominique Certon [1], Massimo Camarda [2], Nicolò Piluso [3], Laurent Colin [1], Flavien Barcella [1] and Daniel Alquier [1,*]

1 GREMAN UMR-CNRS 7347, Université de Tours, INSA Centre Val de Loire, 37071 Tours, France; jaweb.benmessaoud@univ-tours.fr (J.B.M.); jean-francois.michaud@univ-tours.fr (J.-F.M.); dominique.certon@univ-tours.fr (D.C.); laurent.colin@univ-tours.fr (L.C.); flavien.barcella@univ-tours.fr (F.B.)

2 Paul Scherrer Institute, ODRA/116, 5232 Villigen, Switzerland; massimo.camarda@psi.ch

3 STMicroelectronics, Stradale Primosole, 50, 95121 Catania, Italy; nicolo.piluso@st.com

* Correspondence: daniel.alquier@univ-tours.fr; Tel.: +33-2-47-36-67-45

Received: 1 October 2019; Accepted: 19 November 2019; Published: 21 November 2019

Abstract: The stress state is a crucial parameter for the design of innovative microelectromechanical systems based on silicon carbide (SiC) material. Hence, mechanical properties of such structures highly depend on the fabrication process. Despite significant progresses in thin-film growth and fabrication process, monitoring the strain of the suspended SiC thin-films is still challenging. However, 3C-SiC membranes on silicon (Si) substrates have been demonstrated, but due to the low quality of the SiC/Si heteroepitaxy, high levels of residual strains were always observed. In order to achieve promising self-standing films with low residual stress, an alternative micromachining technique based on electrochemical etching of high quality homoepitaxy 4H-SiC layers was evaluated. This work is dedicated to the determination of their mechanical properties and more specifically, to the characterization of a 4H-SiC freestanding film with a circular shape. An inverse problem method was implemented, where experimental results obtained from bulge test are fitted with theoretical static load-deflection curves of the stressed membrane. To assess data validity, the dynamic behavior of the membrane was also investigated: Experimentally, by means of laser Doppler vibrometry (LDV) and theoretically, by means of finite element computations. The two methods provided very similar results since one obtained a Young's modulus of 410 GPa and a residual stress value of 41 MPa from bulge test against 400 GPa and 30 MPa for the LDV analysis. The determined Young's modulus is in good agreement with literature values. Moreover, residual stress values demonstrate that the fabrication of low-stressed SiC films is achievable thanks to the micromachining process developed.

Keywords: 4H-SiC; bulk micromachining; electrochemical etching; circular membrane; bulge test; vibrometry; mechanical properties; Young's modulus; residual stress; FEM

1. Introduction

Silicon carbide (SiC) is recognized as an excellent material for power applications due to notable electrical properties but SiC is also an outstanding candidate for microelectromechanical systems (MEMS) thanks to its distinguished chemical and mechanical properties [1]. Therefore, sensors that are able to detect temperature, gas, and pressure in aerospace or the automotive field are potential applications of SiC-based devices [2–4]. The cubic 3C-SiC polytype is favoured for MEMS applications as it can be epitaxially grown on silicon (Si) substrates and thus offers a low-cost solution for SiC-based MEMS development coupled with the conventional Si technologies. Unfortunately, the lattice mismatch and thermal expansion coefficient difference between the epitaxied 3C-SiC film and the Si substrate lead to high residual stress after deposition and cooling steps [5]. Generally, 3C-SiC suspended films exhibit an important level of stress, typically more than 100 MPa which consequently affects the film mechanical reliability. In order to overcome the difficulties related to the heteroepitaxy, an alternative solution to achieve suspended films, based on SiC, is using the electrochemical etching (ECE) of 4H-SiC wafers. Until now, very few papers deal with the fabrication of 4H-SiC MEMS structures. Thus, Nida et al. proposed a promising technique employing a highly selective etching of 4H-SiC homoepitaxy films [6]. The etching process stops at the interface between n^+ 4H-SiC substrate and n^- 4H-SiC epilayer, enabling the fabrication of freestanding thin films. Moreover, the crystalline quality of the epilayer, and so, the suspended thin-film, is not affected by the process [7]. This method could pave the road for the fabrication of novel SiC-based detectors [6]. However, designing such original film for MEMS applications requires knowing the mechanical properties of the epilayers. Several popular approaches exist to monitor the static behavior of free-standing films: Curvature measurements [8,9], beam-bending testing [10,11], Raman spectroscopy [12,13], nanoindentation [14], and bulge test [15,16]. Moreover, dynamic techniques based on the resonance frequency determination of thin film are also intensively used [17,18]. In order to determinate the mechanical properties (Young's modulus and residual stress values) of the 4H-SiC film, this study aims to evaluate two experimental techniques: The bulge test and the vibrating method.

The static deflection analysis of a circular shape membrane submitted to high external pressure (up to 4 bars) was implemented. In other words, this method enables to discriminate easily residual stress effects from plate behavior and so, to measure simultaneously residual stress and Young's modulus values. In addition, the static load-deflection curve of a circular film is described by a simple analytical expression, that depends on two fitting parameters only, easy to determine from experimental data [19]. However, several relations between these fitting parameters and the mechanical properties of the film were proposed in the literature, leading to a non-unique solution for the mechanical property value determination [20]. Therefore, a complementary method, based on the membrane vibration study, was used to help in the determination of the thin-film mechanical properties. For that, the dynamic behavior of the film was measured by means of laser Doppler vibrometry (LDV), i.e., resonance mode investigations. An inverse problem approach, based on finite element computations, was also implemented to determine the mechanical properties of the fabricated film.

This study focused on the characterization of a circular 4H-SiC freestanding film. After the membrane preparation description, the methods and experiments are presented. The Young's modulus and the residual stress of the membrane are then extracted using the bulge test results. Finally, combining the resonance frequency measurements with the finite element model allowed refining the residual stress value. This last parameter seems to have an important influence on the membrane mechanical behavior.

2. Materials and Methods

2.1. Sample Preparation

The 4H-SiC membrane was fabricated starting from a wafer with homo-epitaxial layer grown on 375 μm thick 10^{18} cm^{-3} n-type substrate (supplied from CREE®, Durham, NC, USA). The epitaxial layer is around 9 μm thick 10^{13} cm^{-3} n-type. The substrate removal was obtained by electrochemical etching. ECE is an oxidation/oxide-removal process obtained by dipping the SiC wafer in a hydrofluoridric acid-based solution and electrically supply holes for the oxidation through a 100 nm aluminium back metal contact [6,21–23]. The process is capable of removing highly doped ($\geq 10^{18}$ cm^{-3}) p-type and n-type layers but is selective towards low-doped n-type layers (selectivity > 1000:1 with respect to the 5×10^{13} cm^{-3} doped n^- layer). Hence, this process allows the full removal of the highly doped substrate and the local release of the epitaxial layers. The realized membranes have thicknesses and uniformities determined by the epitaxial layer. The 4H-SiC suspended film with circular shape was fabricated at the Paul Scherrer Institute [6]. In addition, the circular shape is the most appropriate geometry in order to study the effects related to the internal stress of a film. Indeed, the stress is equi-biaxial, so the loading will not produce any discontinuity.

A Lext OLS4100 Laser Scanning Microscope (LSM), from Olympus Corporation (Shinjuku-ku, Tokyo, Japan), was used in order to observe the full membrane using the stitching mode. Figure 1a shows the image of the membrane shape after ECE process. One can observe a slight membrane asymmetric geometry, which is due to the fabrication process, i.e., to a non-uniform under-etching of the substrate with respect to the etching mask. Therefore, we assumed that the membrane can be reasonably assimilated as a circle. The diameter measurements were performed in several directions and an average diameter value was evaluated at 4.5 mm. Thickness determination was performed using a Strata DB235 Focused Ion Beam (FIB), from Thermo Fisher Scientific (Waltham, MA, USA). To do that, we injected silver paste through the cavity of the membrane in order to fix and stiffen the suspended film to prevent any vibrations during the observations. After that, the FIB cross-section images were carried out in order to obtain a direct measurement of the film thickness as shown in Figure 1b. Thus, a membrane thickness of 8.8 μm was measured and a thickness variation of ± 0.2 μm was evaluated. As expected, the thin film has a well-controlled thickness because the 4H-SiC epilayer acts as an etching stop.

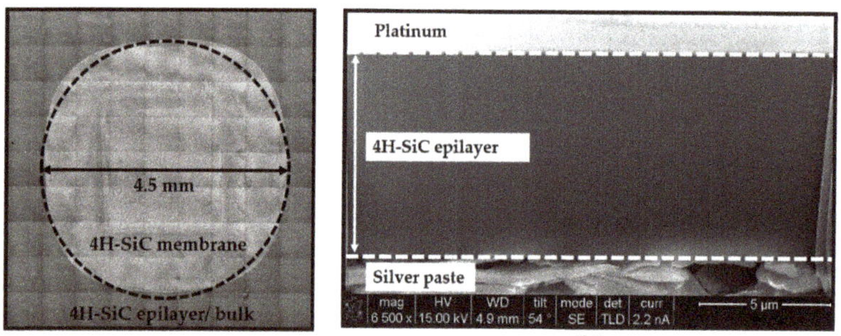

(a) (b)

Figure 1. (**a**) Laser Scanning Microscope (LSM) image of the 4H-SiC membrane using stitching mode, obtained after the electrochemical etching (ECE) process. Circular insert added to show that the 4H-SiC membrane can be assimilated as a circle. (**b**) Focused ion beam (FIB) cross-section image allowing the membrane thickness determination.

2.2. Circular Membrane Deflection under Uniform Pressure

The bulge test method consists in submitting a membrane to uniform external pressure in order to observe its load-deflection behavior. At low value, the mechanical return forces of the membrane are mainly due to the residual stress. Whereas, when the pressure value increases, the return forces from plate stiffness govern the displacement. The deflection of a clamped membrane is measured according to the applied pressure. The most common pressure-deflection relationship of a pre-stressed circular membrane can be expressed as [24]:

$$P(h) = C_1 \frac{t\sigma_0}{a^2} h + C_2 \frac{t}{a^4}\left(\frac{E}{1-v}\right)h^3 = Ah + Bh,^3 \quad (1)$$

where P is the applied pressure, t is the membrane thickness, σ_0 is the residual stress, a is the membrane radius, h is the maximum bulge deflection at the centre of the membrane, E is the Young's modulus, and v is the Poisson's ratio of the 4H-SiC thin-film. C_1 and C_2 are dimensionless constants, which depend on the membrane shape. Note that C_2 is also a function of the Poisson's ratio. A schematic representation of a circular diaphragm is shown in Figure 2. By fitting the applied pressure as a function of the measured deflection, A and B coefficients can be estimated, leading to the determination of the Young's modulus and the residual stress.

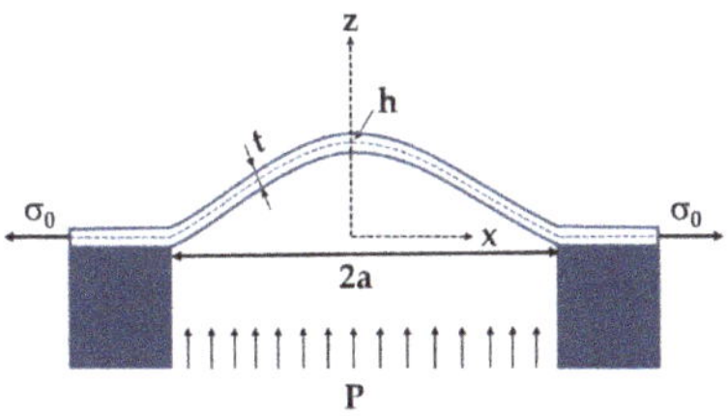

Figure 2. Schematic cross-sectional membrane with initial in-plane tension under uniform pressure P.

The models describing the load-deflection behavior of a circular plate as a function of the pressure have been extensively discussed. Several authors examined the large deflection behavior for the pure plate case, originally described by Nádai and Way [25,26]. Beams was the first to report an experimental model using bulge test to measure the mechanical properties of thin films deposited on substrates [27]. For a circular membrane, Beams determined C_1 and C_2 values, 4 and 8/3, respectively. A more accurate numerical solution indicated that C_2 can be expressed as (8/3) × (1.015 − 0.247υ). Table 1 summarizes the reported value of C_1 and C_2 for circular suspended films from literature. For comparison purposes, C_2 values assuming υ = 0.25 are also listed.

Table 1. C_1 and C_2 coefficients for circular membranes.

Models	C_1	C_2	$C_2(\upsilon = 0.25)$	Approach
Lin [28]	4.0	$(7 - \upsilon)/3$	2.25	Energy minimization
Beams [27]	4.0	8/3	2.67	Spherical cap
Small et al. [29]	4.0	$(8/3) \times (1 - 0.241 \times \upsilon)$	2.51	Finite Element Method
Pan et al. [30]	4.0	$(8/3)/(1.026 + 0.233 \times \upsilon)$	2.46	Finite Element Method
Hohlfelder et al. [31]	4.0	$(8/3) \times (1.015 - 0.247 \times \upsilon)$	2.54	Number approximation
Timoshenko et al. [32]	-	$(8/3) \times 0.976/(1 + \upsilon)$	2.08	Energy minimization

2.3. Membrane Vibration

2.3.1. Analytical Description

The natural frequencies and mode shapes of membranes are a function of the structural properties and boundary conditions. Let us consider a circular membrane with a uniform thickness t and firmly clamped at its periphery. If the displacement of the membrane along the z-axis is small compared to the diameter, the stress state is isotropic and the effect of the fluid (air) is negligible [19]. Therefore, in small deflection case, an ideal membrane is described by assuming that the residual stress is higher than the bending rigidity. Thus, the membrane can support only tensile loads [33]. The mode shapes are described by two spatial coordinates and identified by the nodal line (m, n). Here, m and n are the index relative to the modal diameter lines and the number of nodal circle lines, respectively [33]. The natural frequencies of vibration for a circular membrane can be described as:

$$f_{(m,\,n)} = \frac{\alpha_{mn}}{2\pi a}\frac{\sqrt{\sigma_0}}{\rho}, \quad (2)$$

where a is the membrane radius, σ_0 is the residual stress, ρ is the density of the 4H-SiC thin-film and α_{mn} values are derived from the roots of the first order Bessel functions. Zeros of the Bessel functions can be computed and tabulated [17].

2.3.2. Finite Element Method Approach

A behavioral model of the membrane, based on the finite element method (FEM), was implemented to theoretically determine the mechanical properties of the fabricated film. Therefore, FEM calculations were used as virtual experiments to determine the natural frequencies and mode shapes of the membrane. The numerical model was built with the commercial COMSOL Multiphysics software package using eigenfrequency and solids mechanics modules. In the simulation procedure, a three-dimensional pre-stressed circular membrane was created. Perfectly clamped boundary conditions were used corresponding to a null mechanical displacement at the outer edge, in x, y, and z directions. The mesh settings were defined using the physics-controlled mesh option proposed by COMSOL Multiphysics, giving hence the following parameters: Tetrahedral shape elements and 8767 meshing nodes. These parameters were validated by comparing theoretical resonance frequency values of circular membranes to values obtained from FEM, a difference lower than 1% was obtained.

The residual stress and Young's modulus values extracted from the bulge test measurements were used as input parameters.

2.4. Load-Deflection Measurements

The bulge test setup is fully discussed in the literature, mainly for improving the methodology and the technique accuracy [16,29,34]. Indeed, the deflection measurement errors can lead to an over or under estimation of the mechanical properties. Consequently, this method requires paying special attention to the chip preparation and deflection measurements. The experimental apparatus is schematically presented in Figure 3a. The sample was mounted on a 3 × 3 cm^2 printed circuit board (PCB) holder with a drilled hole in the centre. Several studies have highlighted the importance of the bonding step for the reliability of the deflection measurements. The most common approach to fix the sample is to add adhesive around its edges. In such case, Jayaraman et al. observed that the sample moved during the measurement for a pressure up to 2.8 bars [35]. Mitchell et al. proposed a multi-step bonding method to seal and constrain the sample to the chuck without any displacement of the substrate [34]. Inspired by this method, we deposited an Ablebond 84-3J epoxy adhesive on the PCB and mounted the sample on it. Then, we applied the adhesive around all the edges of the sample to seal and prevent air leakage. Lastly, an annealing step of 1 h at 150 °C was carried out. For the bulge testing, the chip was placed in an airtight square cell, drilled on two lateral faces, in order to inject and measure the air pressure. The membrane is pressurized through its cavity while the front face remains at the atmospheric pressure. So, the sample was characterized under differential pressures, between 0.04 and 4 bars. Pressure regulation and measurements were carried out using both pressure controller and sensors, operating in the range of 0 to 4 bars.

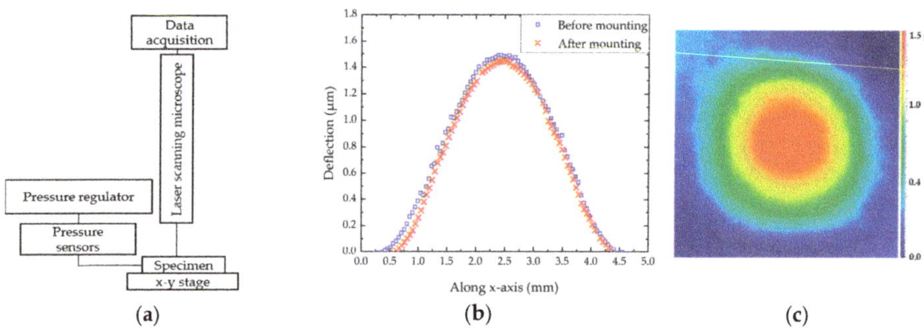

Figure 3. (**a**) Schematic of bulge test apparatus; (**b**) deflection of the circular 4H-SiC membrane, before and after, sample mounting; (**c**) typical topography used to measure the diaphragm deflection with LSM measurement.

Classically, these measurements are performed with a setup integrating both a laser interferometer to detect the membrane deflection, an optical system to observe interference fringes and data processing software [14–18]. Despite the highest resolution of the interferometric method, the experiences had shown that the interference fringes are often not well defined at very small displacements [29]. Moreover, the measurement is often focused in the centre of the membrane, preventing the observation of the deflection profile. Acquisition of load-deflection data is significantly improved using LSM. With its confocal optical system, an LSM detects in-focus reflections from a single specified focal plane along the z-axis [36]. This allows the extraction of the 3D deflection profile of the membrane under an applied pressure.

The deflection h of the membrane (at $P = P_{atm}$) was measured before and after the chip preparation. Values of h are close, around 1.5 µm. So, the impact of the stress induced by the sample preparation seems to be negligible as shown in Figure 3b. The deflection profile was recorded at each stabilized pressure level. A scanning area of 4700 × 4700 µm^2 was defined in the centre of the membrane. Thus, we obtained a mapping including the maximum deflection point in the centre and also the edges of the membrane as presented in Figure 3c.

2.5. Dynamic Behavior Measurements

The natural frequencies of thin film can be measured by several methods. Each technique integrates mechanical excitation setup and optical system. In this study, we used a piezoelectric actuation and a laser system for measuring the film deformation amplitude. The sample was glued using silver paste onto a lead zirconate titanate (PZT) disk and excited by a periodic burst signal at a voltage range between 0.1–10 Vpp. An MSA-500 scanning laser Doppler vibrometry (LDV), from Polytec GmbH (Waldbronn, Germany), was used in order to experimentally investigate the dynamic behavior of the membrane through the identification of its resonance modes. Measurements were performed in air. The diameter of the diaphragm exceeded the aperture angle of the MSA lenses. Consequently, for one acquisition, an area of 890 × 660 µm^2 is measured using the 20× objective. Thus, a stitching method was developed in order to scan the whole membrane surface allowing the determination of its vibrations mode shape. Among all the observed modes, we focused our attention on the first six out-of-plane vibration modes. As an example, Figure 4 shows different measured mode shapes of the 4H-SiC circular membrane.

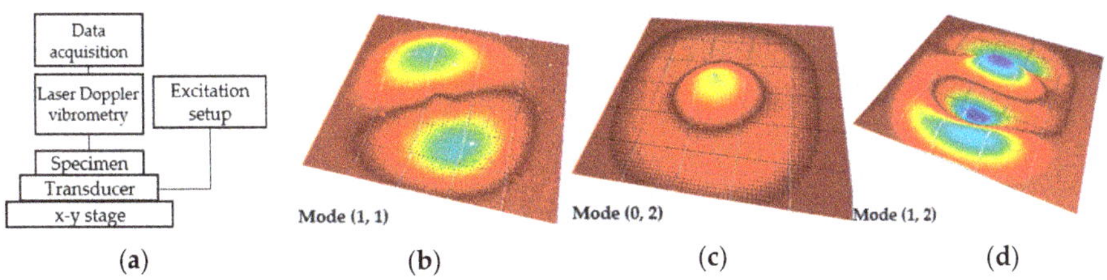

Figure 4. (**a**) Schematic diagram of the resonance frequency method. Vibration mode shapes measured using laser Doppler vibrometry for (**b**) (1, 1); (**c**) (0, 2); and (**d**) (1, 2) modes.

3. Results and Discussion

3.1. Bulge Test Results

As previously explained, the Young's modulus and the residual stress can be determined using the least square fit of Equation (1). The load-deflection response at the membrane centre h versus P is shown in Figure 5.

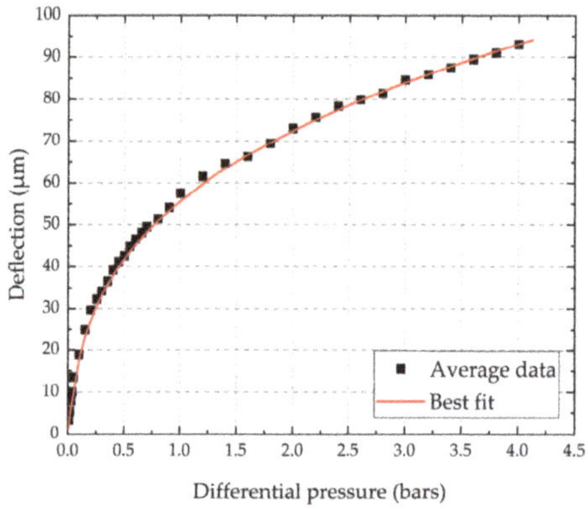

Figure 5. Dot line: Bulge test results for 4H-SiC diaphragm. Solid line: Theoretical fit.

For the calculations, we fixed the Poisson's ratio value at 0.25, which is the most common value used in the literature. It has been pointed out by several authors that the influence of υ in the bulge test is negligible [34,37]. Indeed, our calculations confirm that a variation of υ from 0.2 to 0.3 impacts the Young's modulus value for less than 10%. The parameters used for fitting the data are listed in Table 2.

Table 2. Parameters used for both bulge test and finite element method (FEM) calculations.

Parameters	Values
A (Pa/m)	3.0×10^8
B (Pa/m^3)	4.6×10^{17}
Density (kg/m^3)	3210
Poisson's ratio	0.25
Membrane radius (µm)	2250
Membrane thickness (µm)	8.8

Table 3 reports the determined values of E and σ_0. The calculated Young's modulus values are scattered, depending on the model used. In fact, the main difference between these models is the expression of C_2. Beams was the first to report a value for this dimensionless coefficient, using the spherical cap model based on a very simple approximation of the real case. However, it can lead to an under estimation of the mechanical property determination [29,38]. Therefore, the models proposed by Pan et al. and Small et al. led to close Young's modulus values, which seems to be normal as both models were adjusted from finite element calculations [20]. Moreover, using the numerical solution proposed by Hohlfelder, we obtained almost the same Young's modulus value. Mitchell et al. explained that the difference in the governing equation, which results in measured values, could vary by as much as 20% [34]. In any case, the calculated Young's modulus values are in reasonably good agreement with the published results in the literature for silicon carbide thin films. The residual stress value is determined using the linear term in Equation (1). In comparison with E, the stress σ_0 seems to be less dependent on the model used since C_1 is considered as constant.

Table 3. Bulge test results depending on the models used.

Models	E (GPa)	σ_0 (MPa)
Lin [28]	452	41
Beams et al. [27]	380	41
Small et al. [29]	405	41
Pan et al. [30]	413	41
Hohlfelder [31]	400	41
Mean value	410	41

3.2. Vibrometry Results

In this study, we focused our purpose on the observed six-first vibration modes, which are clearly identified, even if the vibration amplitudes were small. The displacement spectrum of the 4H-SiC circular membrane is shown in Figure 6. The interference frequencies due to the piezoelectric excitation were also measured. The resonance peaks for asymmetric modes (1, 1), (2, 1), (3, 1), and (1, 2) seem to be splitted with lower magnitude peaks. The fabrication process led to a geometric asymmetry of the diaphragm, as previously shown in Figure 1, resulting in the creation of non-degenerated modes in asymmetric vibration modes [39,40]. Moreover, Fartash et al. reported that the presence of an anisotropic tension, due to internal stress or tension after mounting the sample, could also cause the splitting of degenerated modes. Thus, this phenomenon could shift the peaks of asymmetric vibration modes [41].

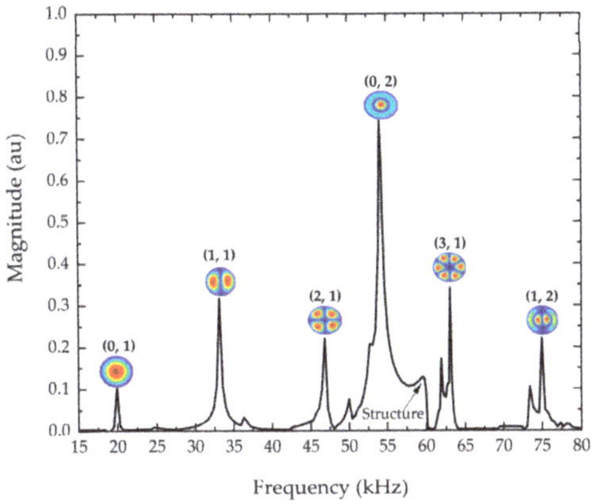

Figure 6. Measured spectrum of vibration of the 4H-SiC membrane associated with the corresponding mode shapes.

3.3. Finite Element Computations

The Young's modulus and the residual stress values determined with the bulge test method were used to calculate the resonance frequencies from the implemented FEM model. Physical and geometrical parameters, already presented in Table 2, were used for the calculations. Figure 7a shows the simulated and measured resonance frequencies, depending on the vibrating mode shapes. Considering the mechanical properties extracted from the bulge test, the calculated resonance frequency values are slightly overestimated, with a difference of 15% compared to the measured values. Moreover, the Young's modulus value seems to have a small, but not negligible, influence on the resonance frequency values. In addition, the membrane shape is not a perfect circle, as already presented in Figure 1a. Therefore, the same calculations were performed using a membrane radius of 2.35 mm, corresponding to the higher measured radius. The obtained results were very close to the preceding one, using a = 2.25 mm. Thus, we assumed that the radius variation has a negligible influence on the presented results. It is important to note that all the calculations were performed for a single residual stress value, i.e., 41 MPa. In order to evaluate the impact of the residual stress, we extended the calculation by changing σ_0 between 15 and 45 MPa. Figure 7b presents the simulations results for the first resonance frequency. The couple (E, σ_0) that provides the best fit with the measured resonance frequency was obtained for E = 400 GPa and σ_0 = 30 MPa. Consequently, we used these mechanical parameters to calculate the five other resonance frequencies. We reported the results in Figure 7a. The resonance frequencies obtained by FEM simulations using σ_0 = 30 MPa seem to be in good agreement with the measured resonance frequencies. Thus, it can be assumed that the residual stress value mainly governs the mechanical behavior of the membrane.

However, the residual stress value extracted from bulge test and FEM simulations are slightly different, even if the difference remains minor. This can be explained by the deflection measurements in the low-pressure range. The determination of the residual stress is allowed with the linear term of Equation (1). Thus, an error in the pressure or deflection measurements could lead to an over or under estimation of the residual stress value.

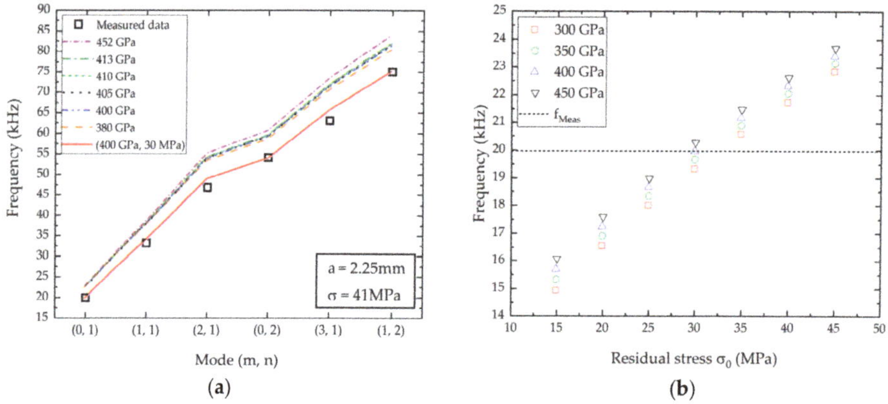

Figure 7. (**a**) Dashed lines: Computed resonance frequencies obtained with FEM calculations using the bulge test results. Solid line: Adjusted FEM calculations with the couple (E, σ_0). Square symbols: Measured resonance frequencies determined with the vibrometry method; (**b**) calculated resonance frequency depending on E and σ_0. Dot line: Measured resonance frequency for the (0, 1) mode. Symbol lines: Calculated resonance frequencies depending on the residual stress and Young's modulus values.

3.4. Etching Profile Determination

As previously explained, we considered, for our model, that the membrane was firmly clamped at its periphery. This property is essential as an undercut or overcut at the diaphragm boundary could shift the resonance frequencies, and so, could affect the accuracy of the mechanical properties determination.

Most of the time, designing freestanding diaphragms based on 3C-SiC material requires etching the Si substrate by using an anisotropic wet etchant, such as potassium hydroxide. This technique is very suitable to create thin suspended square and rectangular membranes but does not allow the achievement of vertical sidewalls. In addition, the realization of circular membrane geometry is more complex, that's why, dry etching using plasma is recommended [39]. With this method, the Si wafer/3C-SiC epilayer interface can act as an etch-stop to define a 3C-SiC membrane. For 4H-SiC, it is much trickier. Indeed, plasma etching of 4H-SiC can also be used to define vertical sidewalls [42]. However, in this case, there is no etching-stop layer. Thus, the realization of a complete well-controlled 4H-SiC membrane is quite impossible. While, thanks to the ECE method applied in this paper, it is achievable. Nonetheless, considering this unusual method, it is mandatory to explore the etching profile. To do that, we observed the sample cross-section by means of LSM method, as shown in Figure 8.

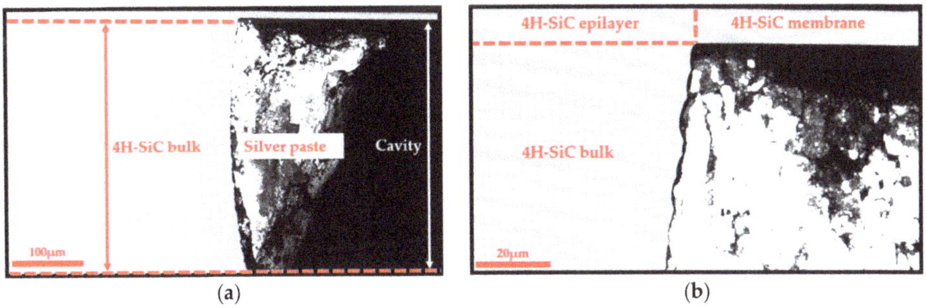

Figure 8. (**a**) LSM cross-section image of the etching profile; (**b**) LSM image of the membrane-undercut boundary.

This figure clearly highlights the anisotropic etching of the 4H-SiC wafer and the thickness homogeneity of the 4H-SiC membrane. Moreover, the assumption of a firmly clamped membrane is clearly demonstrated. It confirms that the ECE method is helpful to achieve well-defined 4H-SiC membranes.

4. Conclusions

The mechanical properties of a 4H-SiC circular membrane were investigated through the determination of its Young's modulus and residual stress. Two methods were implemented and compared. The first one was based on bulge test, where the static behavior of the membrane submitted to a high external pressure was monitored. The second method was based on LDV measurements, where the dynamic mechanical response of the membrane was investigated. An inverse problem approach, based on finite element method, was implemented in order to combine the static and dynamic results, allowing the determination of the mechanical properties of the 4H-SiC circular membrane. The two methods provided very similar results since one obtained a Young's modulus of 410 GPa and a residual stress value of 41 MPa from bulge test against 400 GPa and 30 MPa for the LDV analysis. The calculated Young's modulus is in good agreement with literature values for SiC thin film. Moreover, the process allows the full removal of the highly doped 4H-SiC substrate and the local release of the epitaxial layer, thus realizing membranes with thickness uniformity determined by the epitaxial layer growth. Consequently, a new range of MEMS can be developed using 4H-SiC based-material.

Author Contributions: M.C. conceived the 4H-SiC circular membrane; J.B.M. performed the experiments; D.C. provided simulation tools; N.P., F.B. and L.C. designed the experiments; J.B.M., J.-F.M., M.C., and D.A. participated in the writing of the original draft; J.B.M. wrote the paper.

Funding: This research received no external funding.

Acknowledgments: J.B.M. wishes to thank "Region Centre Val de Loire" and "Conseil Géneral d'Indre et Loire" for the financial support of this work, in the framework of his PhD fellowship.

Conflicts of Interest: The authors declare no conflict of interest.

References

1. Sarro, P.M. Silicon carbide as a new MEMS technology. *Sens. Actuators A Phys.* **2000**, *82*, 210–218. [CrossRef]
2. Elyassi, B.; Sahimi, M.; Tsotsis, T.T. Silicon carbide membranes for gas separation applications. *J. Membr. Sci.* **2007**, *288*, 290–297. [CrossRef]
3. Mehregany, M.; Zorman, C.A.; Rajan, N.; Wu, C.H. Silicon carbide MEMS for harsh environments. *Proc. IEEE* **1998**, *86*, 1594–1609. [CrossRef]
4. Michaud, J.-F.; Portail, M.; Alquier, D. 3C-SiC—From Electronic to MEMS Devices. In *Advanced Silicon Carbide Devices and Processing*; InTech: Philadelphia, PA, USA, 2015; p. 33. ISBN 978-953-51-2168-8.

5. Zielinski, M.; Leycuras, A.; Ndiaye, S.; Chassagne, T. Stress relaxation during the growth of 3C-SiC/Si thin films. *Appl. Phys. Lett.* **2006**, *89*, 2004–2007. [CrossRef]
6. Nida, S.; Tsibizov, A.; Ziemann, T.; Woerle, J.; Moesch, A.; Schulze-Briese, C.; Pradervand, C.; Tudisco, S.; Sigg, H.; Bunk, O.; et al. Silicon carbide X-ray beam position monitors for synchrotron applications. *J. Synchrotron Radiat.* **2019**, *26*, 28–35. [CrossRef]
7. Dahal, R.; Chowdhury, S.; Hitchcock, C.; Chow, T.P.; Bhat, I.B. Fabrication of thick free-standing lightly-doped n-type 4H- SiC wafers. *Mater. Sci. Forum* **2017**, *897*, 379–382. [CrossRef]
8. Janssen, G.C.A.M.; Abdalla, M.M.; van Keulen, F.; Pujada, B.R.; van Venrooy, B. Celebrating the 100th anniversary of the Stoney equation for film stress: Developments from polycrystalline steel strips to single crystal silicon wafers. *Thin Solid Films* **2009**, *517*, 1858–1867. [CrossRef]
9. Feng, X.; Huang, Y.; Rosakis, A.J. On the Stoney Formula for a thin film/substrate system with nonuniform substrate thickness. *J. Appl. Mech.* **2007**, *74*, 1276–1281. [CrossRef]
10. Bazin, A.-E.; Chassagne, T.; Alquier, D.; Zielinski, M.; Jiao, S.; Michaud, J.-F.; Portail, M. Micromachining of thin 3C-SiC films for mechanical properties investigation. *MRS Proc.* **2010**, *1246*, 3–9. [CrossRef]
11. Fang, W.; Wickert, J.A. Determining mean and gradient residual stresses in thin films using micromachined cantilevers. *J. Micromech. Microeng.* **1996**, *6*, 301–309. [CrossRef]
12. Yang, S.; Miyagawa, R.; Miyake, H.; Hiramatsu, K.; Harima, H. Raman scattering spectroscopy of residual stresses in epitaxial AlN films. *Appl. Phys. Express* **2011**, *4*, 10–13. [CrossRef]
13. Piluso, N.; Anzalone, R.; Camarda, M.; Severino, A.; D'Arrigo, G.; La Via, F. Stress fields analysis in 3C-SiC free-standing microstructures by micro-Raman spectroscopy. *Thin Solid Films* **2012**, *522*, 20–22. [CrossRef]
14. Oliver, W.C.; Pharr, G.M. Measurement of hardness and elastic modulus by instrumented indentation: Advances in understanding and refinements to methodology. *J. Mater. Res.* **2004**, *19*, 3–20. [CrossRef]
15. Xiang, Y.; Chen, X.; Vlassak, J.J. Plane-strain bulge test for thin films. *J. Mater. Res.* **2005**, *20*, 2360–2370. [CrossRef]
16. Zhou, W.; Yang, J.; Sun, G.; Liu, X.; Yang, F.; Li, J. Fracture properties of silicon carbide thin films by bulge test of long rectangular membrane. *J. Microelectromech. Syst.* **2008**, *17*, 453–461. [CrossRef]
17. Maden, M.A.; Jagota, A.; Mazur, S.; Farris, R.J. Vibrational technique for stress measurement in films: I, ideal membrane behavior. *J. Am. Ceram. Soc.* **1994**, *77*, 625–635. [CrossRef]
18. Yaralioglu, G.G.; Ergun, A.S.; Bayram, B.; Marentis, T.; Khuri-Yakub, B.T. Residual stress and Young's modulus measurement of capacitive micromachined ultrasonic transducer membranes. *Proc. IEEE* **2001**, *2*, 953–956. [CrossRef]
19. Vlassak, J.J.; Nix, W.D. A new bulge test technique for the determination of Young's modulus and Poisson's ratio of thin films. *J. Mater. Res.* **1992**, *7*, 3242–3249. [CrossRef]
20. Malhaire, C. Comparison of two experimental methods for the mechanical characterization of thin or thick films from the study of micromachined circular diaphragms. *Rev. Sci. Instrum.* **2012**, *83*, 055008. [CrossRef]
21. Watanabe, N.; Kimoto, T.; Suda, J. Fabrication of electrostatic-actuated single-crystalline 4H-SiC bridge structures by photoelectrochemical etching. *Proc. SPIE* **2011**, *7926*, 79260B. [CrossRef]
22. Gautier, G.; Biscarrat, J.; Valente, D.; Defforge, T.; Gary, A.; Cayrel, F. Systematic study of anodic etching of highly doped N-type 4H-SiC in various HF based electrolytes. *J. Electrochem. Soc.* **2013**, *160*, 372–379. [CrossRef]
23. Gautier, G.; Cayrel, F.; Capelle, M.; Billoué, J.; Song, X.; Michaud, J.-F. Room light anodic etching of highly doped n-type 4H-SiC in high-concentration HF electrolytes: Difference between C and Si crystalline faces. *Nanoscale Res. Lett.* **2012**, *7*, 367. [CrossRef] [PubMed]
24. Vinci, R.P.; Vlassak, J.J. Mechanical behavior of thin films. *Annu. Rev. Mater. Sci.* **1996**, *26*, 431–462. [CrossRef]
25. Nádai, A. *Die Elastischen Platten*; Springer: Berlin/Heidelberg, Germany, 1925.
26. Way, S. Bending of circular plates with large deflection. *Trans. ASME* **1934**, *54*, 627–636.
27. Beams, J.W. Mechanical properties of thin films of silver. In *Structure and Properties of Thin Films*; John Wiley and Sons: New York, NY, USA, 1959; pp. 183–192.
28. Lin, P. The In-Situ Measurement of Mechanical Properties of Multi-Layer Coatings. Ph.D. Thesis, Massachusetts Institute of Technology, Cambridge, MA, USA, 1990.
29. Small, M.K.; Nix, W.D. Analysis of the accuracy of the bulge test in determining the mechanical properties of thin films. *J. Mater. Res.* **1992**, *7*, 1553–1563. [CrossRef]

30. Pan, J.Y.; Lin, P.; Maseeh, F.; Senturia, S.D. Verification of FEM analysis of load-deflection methods for measuring mechanical properties of thin films. In Proceedings of the IEEE 4th Technical Digest on Solid-State Sensor and Actuator Workshop, Hilton Head Island, SC, USA, 4–7 June 1990; pp. 70–73. [CrossRef]
31. Hohlfelder, R.J.; Vlassak, J.J.; Nix, W.D.; Luo, H.; Christopher, E.D. Blister test analysis methods. *Mater. Res. Soc.* **1995**, *356*, 585–590. [CrossRef]
32. Timoshenko, S.P.; Krieger, S.W. *Theory of Plate and Shells*, 2nd ed.; McGraw-Hill: New York, NY, USA, 1969.
33. Blevins, R.D. *Formulas for Natural Frequency and Mode Shape*; Van Nostrand Reinhold Company: New York, NY, USA, 1979.
34. Mitchell, J.S.; Zorman, C.A.; Kicher, T.; Roy, S.; Mehregany, M. Examination of bulge test for determining residual stress, Young's modulus, and Poisson's ratio of 3C-SiC thin films. *J. Aerosp. Eng.* **2003**, *16*, 46–54. [CrossRef]
35. Jayaraman, S.; Edwards, R.L.; Hemker, K.J. Relating mechanical testing and microstructural features of polysilicon thin films. *J. Mater. Res.* **1999**, *14*, 688–697. [CrossRef]
36. Olympus. Available online: www.olympus-europa.com (accessed on 19 November 2019).
37. Huang, H.; Liu, Y.; Dell, J.M. Determination of mechanical properties of PECVD silicon nitride thin films for tunable MEMS Fabry–Perot optical filters. *J. Micromech. Microeng.* **2005**, *15*, 608. [CrossRef]
38. Hill, R.C. A theory of the plastic bulging of a metal diaphragm by lateral pressure. *Lond. Edinb. Dublin Philos. Mag. J. Sci.* **1950**, *41*, 1133–1142. [CrossRef]
39. Ma, S.; Wang, S.; Iacopi, F.; Huang, H. A resonant method for determining the residual stress and elastic modulus of a thin film. *Appl. Phys. Lett.* **2013**, *103*, 031603. [CrossRef]
40. Deutsch, B.M.; Robinson, A.R.; Felce, R.J.; Moore, T.R. Nondegenerate normal-mode doublets in vibrating flat circular plates. *Am. J. Phys.* **2004**, *72*, 220–225. [CrossRef]
41. Fartash, A.; Schuller, I.K.; Grimsditch, M. Thinfilm modeling for mechanical measurements: Should membranes be used or plates? *J. Appl. Phys.* **1992**, *71*, 4244–4248. [CrossRef]
42. Biscarrat, J.; Michaud, J.F.; Collard, E.; Alquier, D. ICP Etching of 4H-SiC Substrates. *Mater. Sci. Forum* **2013**, *740–742*, 825–828. [CrossRef]

micromachines

MDPI

Article

An Improved 4H-SiC MESFET with a Partially Low Doped Channel

Hujun Jia *, Yibo Tong, Tao Li, Shunwei Zhu, Yuan Liang, Xingyu Wang, Tonghui Zeng and Yintang Yang

School of Microelectronics, Xidian University, Xi'an 710071, China
* Correspondence: hjjia@mail.xidian.edu.cn; Tel.: +86-137-7212-6387

Received: 8 August 2019; Accepted: 21 August 2019; Published: 23 August 2019

Abstract: An improved 4H-SiC metal semiconductor field effect transistor (MESFET) based on the double-recessed MESFET (DR-MESFET) for high power added efficiency (PAE) is designed and simulated in this paper and its mechanism is explored by co-simulation of ADS and ISE-TCAD software. This structure has a partially low doped channel (PLDC) under the gate, which increases the PAE of the device by decreasing the absolute value of the threshold voltage (V_t), gate-source capacitance (C_{gs}) and saturation current (I_d). The simulated results show that with the increase of H, the PAE of the device increases and then decreases when the value of N_{PLDC} is low enough. The doping concentration and thickness of the PLDC are respectively optimized to be $N_{PLDC} = 1 \times 10^{15}$ cm^{-3} and $H = 0.15$ μm to obtain the best PAE. The maximum PAE obtained from the PLDC-MESFET is 43.67%, while the PAE of the DR-MESFET is 23.43%; the optimized PAE is increased by 86.38%.

Keywords: 4H-SiC; MESFET; simulation; PAE

1. Introduction

With the development of the semiconductor industry, SiC, diamond and GaN, the third-generation semiconductor materials, have become a research hotspot because of their high critical field strength, wide band gap and high carrier saturation rate [1–6]. 4H-SiC is used to manufacture power devices such as MESFETs due to its larger band gap and higher electron mobility compared to those of 3C-SiC and 6H-SiC [7]. Nowadays, the mainstream research direction on 4H-SiC MESFETs is to achieve better output power density by making changes to the device structure [8,9]. However, in order to achieve green development, enabling devices to have higher energy conversion efficiency has become a new central issue of research. In the papers An Improved DRBL AlGaN/GaN HEMT with High Power Added Efficiency [10] and An Improved UU-MESFET with High Power Added Efficiency [11], a higher power added efficiency (PAE) was obtained by balancing the parameters of the devices. The PAE of the improved with an ultrahigh upper gate MESFET (IUU-MESFET) and the double recessed barrier layer (DRBL) AlGaN/GaN HEMT increased 18% and 48%, respectively. In the aforementioned research works, PAE simply replaces the RF output power with the difference between output and input power in the drain efficiency equation. A larger PAE means that a larger output power can be obtained under the same input power. This is crucial for sustainable development.

In this paper, an improved 4H-SiC MESFET with a partially low doped channel (PLDC) is designed and simulated to improve the PAE of the 4H-SiC DR-MESFET [12] using ISE-TCAD and ADS. A partially low doped channel is used to balance the parameters of the device by adjusting the doping concentration and thickness. The key to this structure is to improve the AC/RF characteristics of the device and improve the PAE of the device. This ensures that the device has lower energy consumption at the same output power, which has great significance for RF power amplifier applications. In the second part of this paper, the basic features and simulation process of the PLDC-MESFET are introduced, as are the models used in the simulation. In the third section, the main impact of the PLDC on the parameters and PAE of the device is introduced and the mechanism is discussed. In the fourth section, we conclude that the PLDC is helpful for the improvement of the PAE of the DR-MESFET.

2. Device Structure

The 2D schematic cross-sections of the DR-MESFET and PLDC-MESFET structures are shown in Figure 1a,b, respectively. The difference between the two devices is that the PLDC-MESFET has a partially low doped channel under the gate. The PLDC was realized by high-energy ion implantation and high-temperature annealing processes. It should be noted that the P-type impurity is implanted to compensate for the formation of lightly doped regions [13]. The thickness and the concentration of the PLDC are denoted as H and N_{PLDC}, respectively. The N_{PLDC} was set to 1×10^{17} cm^{-3}, 1×10^{16} cm^{-3} and 1×10^{15} cm^{-3}. The H was set from 0 to 0.25 μm in a step of 0.05 μm.

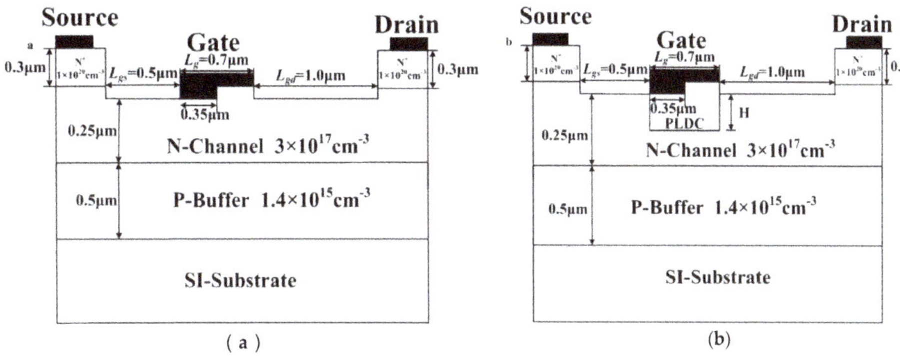

Figure 1. Schematic cross-sections of the (**a**) DR 4H-SiC MESFET, (**b**) partially low doped channel (PLDC) 4H-SiC MESFET.

The main physics models were applied in ISE-TCAD tools simulation [14], including Mobility (Doping Dep, HighFieldSat Enormal), Effective Intrinsic Density (Band Gap Narrowing (OldSlotboom), Incomplete Ionization, Recombination (SRH (Doping Dep) and Auger Avalanche (Eparallel). The criterion of breakdown was Break Criteria {Current (Contact = "gate" Absval = 1e3)}. The main solving model was Coupled {Poisson Electron Hole}. Mobility models were used to solve the phenomenon of the mobility of carriers being degraded by many factors. Recombination models were used to calculating the lifetime of carriers. The Effective Intrinsic Density model was used to calculate the effective band gap. Incomplete Ionization must be considered, as this occurs in the case of aluminum acceptors in silicon carbide. The temperature of the simulations was 300 K. The major parameters of the device measured were saturation current (I_{d}), threshold voltage (V_{t}), gate–source capacitance (C_{gs}) and transconductance (g_{m}). Those parameters are used in ADS to modify the EE_FET3 model. The modified EE_FET3 model and "Load-Pull PAE, Output Power Contours" model [15] were used to measure the PAE of the device under the same bias conditions. The working bias conditions were set as follows: V_{gs} was −8.0 V, V_{ds} was 28 V, RF was 850 MHz and Pavs_dBm was 28 dBm. Keeping the bias condition and changing the parameters obtained from ISE-TCAD, the PAE of the device under different thicknesses and doping concentrations can be calculated as follows [16].

$$\eta(\mathrm{PAE}) = \frac{P_{\mathrm{out}} - P_{\mathrm{in}}}{P_{\mathrm{dc}}} \tag{1}$$

where P_{out} is output power, P_{in} is input power and P_{dc} is DC power.

3. Results and Discussion

3.1. The Effect of Doping Concentration and Thickness On the Device Parameters

As showing in Figure 2, the parameters of the device are greatly affected by the doping concentration (N_{PLDC}) and thickness (H) of the PLDC. The effect of N_{PLDC} and H on V_{t} is shown in Figure 2a. With the decrease of N_{PLDC}, the absolute value of V_{t} decreases obviously. When H increases, the V_{t} overall trend is also decreasing. This is because the changes in N_{PLDC} and H directly control the total carrier concentration in the channel, and V_{t} is proportional to the total carrier. Figure 2b shows the effects of N_{PLDC} and H on C_{gs}. With the decrease of N_{PLDC} and the increase of H, C_{gs} decreases. On the one hand, the PLDC suppresses the under-gate depletion layer extending to the source side, and on the other hand, it reduces the total number of carriers in the channel, thereby reducing the input capacitance of the device. In the Figure 2c, g_{m} increases first and then decreases. The reason for this formation may be that the thinner low doped layer can increase the gate's ability to control the current by inhibiting the diffusion of the depletion layer to some extent. When H is thick enough, the ability of the gate to control the current will be reduced. So, g_{m} decreases. In Figure 2d, I_{dsat} is roughly decreased as H increases and N_{PLDC} decreases. This is mainly caused by the decrease of the channel carrier concentration. When H is 0.25 μm, the parameters exhibit a sharp decrease and the DC characteristic of the device becomes poor. It is indicated by the simulation results that the PLDC-MESFET has smaller values of C_{gs}, g_{m}, V_{t} and I_{dsat} as compared to those of the original device.

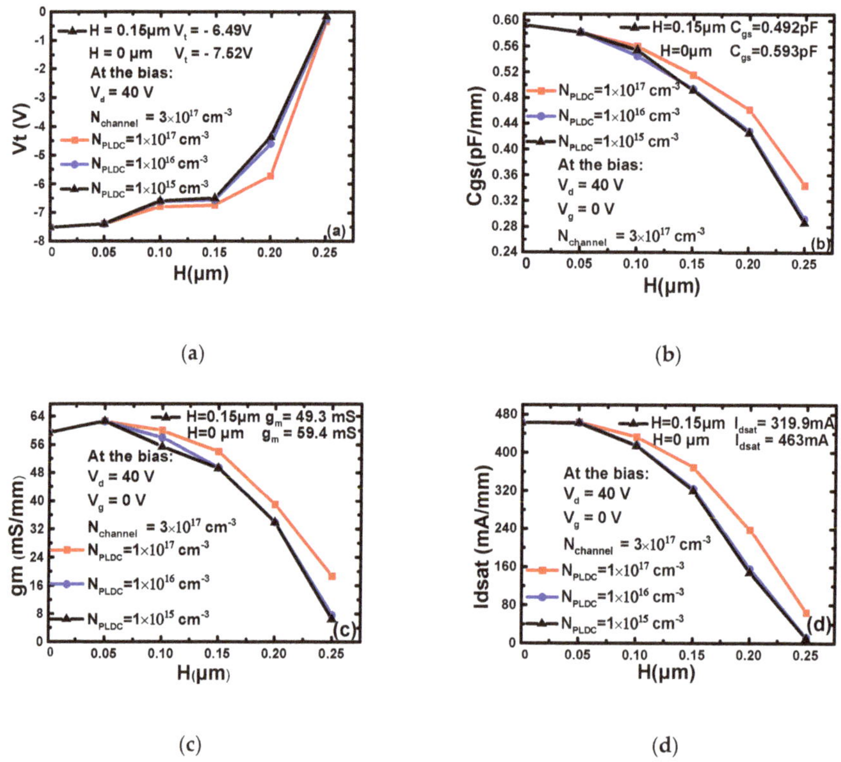

Figure 2. The effect of N_{PLDC} and H on the device parameters: (**a**) V_{t}-N_{PLDC} and H, (**b**) C_{gs}-N_{PLDC} and H, (**c**) g_{m}-N_{PLDC} and H, (**d**) I_{dsat}-N_{PLDC} and H.

3.2. The Influences of Doping Concentration and Thickness on the PAE

The influences of the doping concentration and thickness on the PAE are shown in Figure 3. It can be seen that when H is smaller than 0.20 μm, the PAE of the device increases with the decrease of N_{PLDC}. When H is 0.20 μm and N_{PLDC} is 1×10^{15} cm^{-3} or 1×10^{16} cm^{-3}, the PAE of the device decreases sharply. When H is 0.20 μm and N_{PLDC} is 1×10^{17} cm^{-3}, the PAE of the device increases. When H is 0.25 μm, the simulation results show that the DC characteristics and AC characteristics of the device are poor, and the PAE of these structures is low. The maximum value of the PAE is obtained when the N_{PLDC} is 1×10^{15} cm^{-3}, the H is 0.15 μm. The PAE of the new device is 43.67% while the PAE of the original device is 23.43%. The optimized PAE is increased by 86.38%. The PAE of the IUU-MESFET and DRBL AlGaN/GaN HEMT increase 18% and 48%, respectively. So, the PLDC has a great effect on improving the PAE of the device. In the paper *107 W CW SiC MESFET with 48.1% PAE*, the experimental PAE of the device at 2 W (33 dBm) is close to 25% [17]. The PAE of the DR-MESFET is 23.43% at 0.63 W (28 dBm). This is essentially consistent with the simulation results.

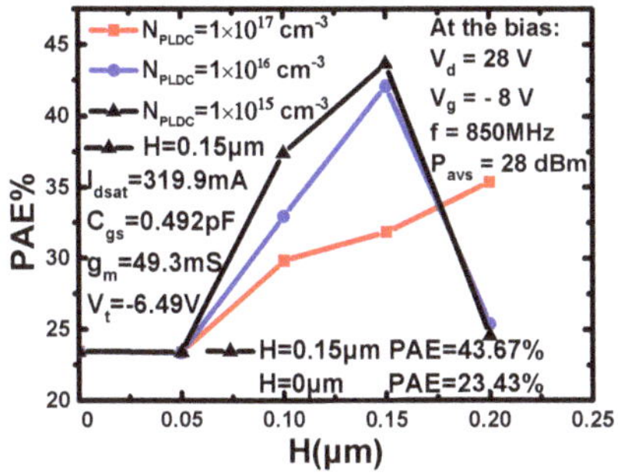

Figure 3. The effects of N_{PLDC} and H on the PAE.

3.3. Mechanism Discussion

Figure 4a,b shows the influence of the parameters on PAE at the same bias when V_{gs} is −8.0 V, V_{ds} is 28 V, RF is 850 MHz and Pavs_dBm is 28 dBm. As shown in Figure 4a, the PAE increases with the increase of V_t when g_m is a constant. When V_t is a constant, the PAE also increases with the increase of g_m. When g_m is between 40 and 60 mS, the PAE of the device has the biggest change. This can be observed by the distance between the two curves. Figure 4b shows the influence of I_{dsat} and C_{gs} on the PAE. With the increase of C_{gs}, the PAE decreases. With the increase of I_{dsat}, the PAE increase. Furthermore, the larger I_{dsat} is, the slower PAE increases.

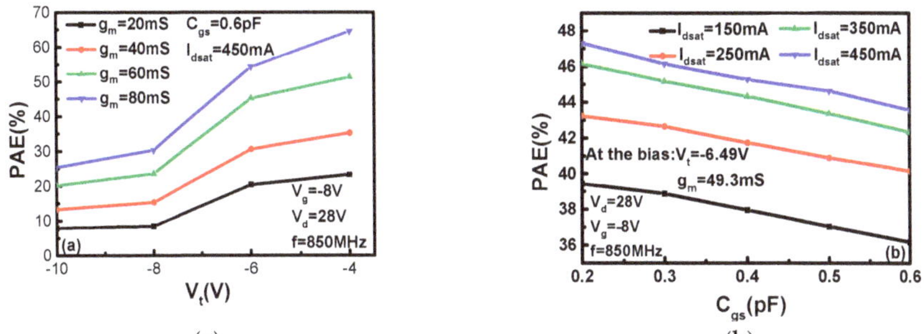

(**a**) (**b**)

Figure 4. The effect of device parameters on PAE: (**a**) PAE-V_t and g_m, (**b**) *PAE*-C_{gs} and I_{dsat}.

From the analysis above, it can be concluded that the smaller the absolute value of V_t, the bigger the PAE, and the smaller the C_{gs}, the bigger the PAE. For g_m, a bigger g_m means a higher current gain, so it a larger output can be obtained under the same input. According to Figure 4a, the PAE is proportional to g_m. This is the reason why the PAE of the device decreases sharply when H is 0.20 μm and N_{PLDC} is 1×10^{15} cm^{-3} or 1×10^{16} cm^{-3}. When H is 0.20 μm and N_{PLDC} is 1×10^{17} cm^{-3}, the PAE of the device increases because g_m is not the key factor compared with V_t, I_{dsat} and C_{gs}. The PAE of the device is decided by the influences of those parameters.

It can be seen that the doping concentration and thickness of the PLDC are optimized to be $N_{PLDC} = 1 \times 10^{15}$ cm^{-3} and $H = 0.15$ μm. Table 1 shows some main parameters of the two devices. It can be seen that the PAE of the PLDC-MESFET is 43.67%, which is higher than the PAE of 23.43% of the DR-MESFET. Compared the two devices, the PLDC-MESFET has a smaller threshold voltage, smaller input capacitance, smaller transconductance and smaller saturation current than the DR-MESFET. The increase of the PAE is influenced by the combination of these parameters. When the absolute value of V_t decreases, the device is easier to turn on and gains a larger output current. So, the output power P_{out} increases and a higher PAE is reached. According to Formula (2) [16], a smaller input capacitance C_{gs} means the device has less energy loss when working in RF (charging and discharging).

$$P_{dyn} = E_{VD} - E_c = \int_0^{\infty} i_{vd}(t)V_d dt - \int_0^{\infty} i_{vd}(t)v_{out} dt = CV_D{}^2 - \frac{CV_D^2}{2} = \frac{CV_D^2}{2} \quad (2)$$

where P_{dyn} is the dynamic power consumption flipped once, E_{VD} is the energy obtained from the power source, E_c is the capacitor stored energy, C is the gate–source capacitor and V_D is the drain voltage. A small C_{gs} also increases the input impedance of the device. Therefore, P_{out} of the device increases and P_{in} decreases. For I_{dsat}, a small I_{dsat} indicates a small P_{out}. Under the influence of these parameters, the device has a big PAE. In there, g_m is sacrificed to obtain a higher PAE. Though a larger g_m is helpful to increase PAE, the influences of the other parameters on PAE are more obvious. So, the maximum value of PAE is 43.67% when N_{PLDC} is 1×10^{15} cm^{-3} and H is 0.15 μm, as obtained by sacrificing some of the DC performances of the device.

Table 1. Comparison of performance parameters of the two structures.

Parameters	DR 4H-SiC MESFET	PLDC 4H-SiC MESFET
I_{dsat} (mA/mm)	448.00	319.90
V_b (V)	125.35	130.20
g_m (mS/mm)	59.30	49.30
V_t (V)	−7.52	−6.49
C_{gs} (pF/mm)	0.59	0.49
PAE (%)	23.43	43.67

4. Conclusions

An improved 4H-SiC MESFET with a partially low doped channel is designed and simulated in this paper to increase the PAE of the device. The results show that the maximum PAE of the PLDC-MESFET is 43.67%, while the PAE of the DR-MESFET is 23.43%; the optimized PAE was increased by 86.38%. A way to design an energy efficient amplifier is proposed in this paper by balancing the parameters of the device. This ensures that the device has lower energy consumption at the same output power, which has great significance for RF power amplifier applications.

Author Contributions: Project administration, H.J.; writing—original draft preparation, Y.T.; writing—review and editing, T.L., S.Z., Y.L., Y.Y.; Data curation, X.W.; Formal analysis, T.Z.

Funding: This work was supported by the National Natural Science Foundation of China (NSFC) under Grant No. 61671343.

Conflicts of Interest: The authors declare no conflict of interest.

References

1. Riaz, M.; Ahmed, M.M.; Munir, U. An improved model for current voltage characteristics of submicron SiC MESFETs. *Solid State Electron.* **2016**, *121*, 54–61. [CrossRef]
2. Nakashima, S.; Harima, H. Raman Investigation of SiC Polytypes. *Phys. Status Solidi Appl. Res.* **2015**, *162*, 39–64. [CrossRef]
3. Kirschman, R. Status of Silicon Carbide (SiC) as a WideBandgap Semiconductor for HighTemperature Applications: A Review. *Solid-State Electron.* **1996**, *39*, 1409–1422.
4. Holmes, J.; Dutta, M.; Koeck, F.A.; Brown, J.; Fox, B.; Hathwar, R.; Johnson, H.; Malakoutian, M.; Saremi, M.; Zaniewski, A.; et al. A 4.5-µm PIN diamond diode for detecting slow neutrons. *Nucl. Instrum. Methods Phys. Res. A* **2018**, *903*, 297–301. [CrossRef]
5. Saremi, M.; Hathwar, R.; Dutta, M. Analysis of the reverse I-V characteristics of diamond-based PIN diodes. *Appl. Phys. Lett.* **2017**, *111*, 043507. [CrossRef]
6. Aminbeidokhti, A.; Dimitrijev, S.; Kumar Hanumanthappa, A. Gate-Voltage Independence of Electron Mobility in Power AlGaN/GaN HEMTs. *IEEE Trans. Electron Devices* **2016**, *63*, 1013–1019. [CrossRef]
7. Aminbeidokhti, A.; Dimitrijev, S.; Han, J. The Power Law of Phonon-Limited Electron Mobility in the 2-D Electron Gas of AlGaN/GaN Heterostructure. *IEEE Trans. Electron Devices* **2016**, *63*, 2214–2218. [CrossRef]
8. Yakimova, R.; Syväjarvi, M.; Iakimov, T.; Jacobsson, H.; Kakanakova-Georgieva, A.; Råback, P.; Janzén, E. Growth of silicon carbide: Process-related defects. *Appl. Surf. Sci.* **2001**, *184*, 27–36. [CrossRef]
9. Jia, H.J.; Yang, Z.H.; Ma, P.M. A novel 4H-SiC MESFET with localized high-doped P-buffer layer. In Proceedings of the IEEE International Conference on Solid-State and Integrated Circuit Technology, Hangzhou, China, 25–28 October 2016; pp. 1086–1088.
10. Jia, H.J.; Zhu, S.W.; Hu, M. An improved DRBL AlGaN/GaN HEMT with high power added efficiency. *Mater. Sci. Semicond. Process.* **2019**, *89*, 212–215. [CrossRef]
11. Jia, H.J.; Hu, M.; Zhu, S.W. An Improved UU-MESFET with High Power Added Efficiency. *Micromachines* **2018**, *9*, 573. [CrossRef] [PubMed]
12. Zhu, C.L.; Tin, C.C.; Zhang, G.H.; Yoon, S.F.; Ahn, J. Improved performance of SiC MESFETs using double-recessed structure. *Microelectron. Eng.* **2005**, *83*, 92–95. [CrossRef]
13. Horiuchi, T.; Homma, T.; Murao, Y.; Okumura, K. An asymmetric sidewall process for high performance LDD MOSFET. *IEEE Trans. Electron Devices* **1994**, *41*, 186–190. [CrossRef]
14. Synopsys Inc. *Dessis User Guide, User Manuals, Version ISE TCAD 10.0*; Synopsys Inc.: Mountain View, CA, USA, 2004.
15. Angilent Technologies. *ADS Documentation, User Manuals, Version ADS2015*; Angilent Technologies: Santa Clara, CA, USA, 2015.
16. Lee, T.H. *The Design of CMOS Radio-Frequency Integrated Circuits*; Cambridge University Press: Cambridge, UK, 1998.
17. Tao, Y.; Bai, S. 107W CW SiC MESFET with 48.1% PAE. In Proceedings of the IEEE International Symposium on Microwave, Antenna, Propagation and EMC Technologies for Wireless Communications, Chengdu, China, 29–31 October 2013.

micromachines

MDPI

Article

Fabrication of a Monolithic Implantable Neural Interface from Cubic Silicon Carbide

Mohammad Beygi [1], John T. Bentley [2], Christopher L. Frewin [3], Cary A. Kuliasha [4], Arash Takshi [1], Evans K. Bernardin [2], Francesco La Via [5] and Stephen E. Saddow [1,2,*]

1 Department of Electrical Engineering, University of South Florida, Tampa, FL 33620, USA
2 Department of Medical Engineering, University of South Florida, Tampa, FL 33620, USA
3 NeuroNexus Technologies, Inc., Ann Arbor, MI 48108, USA
4 Department of Electrical and Computer Engineering, University of Florida, Gainesville, FL 32611, USA
5 CNR Institute for Microelectronics and Microsystems, Catania, Sicily 95121, Italy
* Correspondence: saddow@ieee.org; Tel.: +1-813-974-4773

Received: 13 May 2019; Accepted: 26 June 2019; Published: 29 June 2019

Abstract: One of the main issues with micron-sized intracortical neural interfaces (INIs) is their long-term reliability, with one major factor stemming from the material failure caused by the heterogeneous integration of multiple materials used to realize the implant. Single crystalline cubic silicon carbide (3C-SiC) is a semiconductor material that has been long recognized for its mechanical robustness and chemical inertness. It has the benefit of demonstrated biocompatibility, which makes it a promising candidate for chronically-stable, implantable INIs. Here, we report on the fabrication and initial electrochemical characterization of a nearly monolithic, Michigan-style 3C-SiC microelectrode array (MEA) probe. The probe consists of a single 5 mm-long shank with 16 electrode sites. An ~8 µm-thick p-type 3C-SiC epilayer was grown on a silicon-on-insulator (SOI) wafer, which was followed by a ~2 µm-thick epilayer of heavily n-type (n^+) 3C-SiC in order to form conductive traces and the electrode sites. Diodes formed between the p and n^+ layers provided substrate isolation between the channels. A thin layer of amorphous silicon carbide (*a*-SiC) was deposited via plasma-enhanced chemical vapor deposition (PECVD) to insulate the surface of the probe from the external environment. Forming the probes on a SOI wafer supported the ease of probe removal from the handle wafer by simple immersion in HF, thus aiding in the manufacturability of the probes. Free-standing probes and planar single-ended test microelectrodes were fabricated from the same 3C-SiC epiwafers. Cyclic voltammetry (CV) and electrochemical impedance spectroscopy (EIS) were performed on test microelectrodes with an area of 491 μm^2 in phosphate buffered saline (PBS) solution. The measurements showed an impedance magnitude of 165 kΩ ± 14.7 kΩ (mean ± standard deviation) at 1 kHz, anodic charge storage capacity (CSC) of 15.4 ± 1.46 mC/cm^2, and a cathodic CSC of 15.2 ± 1.03 mC/cm^2. Current-voltage tests were conducted to characterize the p-n diode, n-p-n junction isolation, and leakage currents. The turn-on voltage was determined to be on the order of ~1.4 V and the leakage current was less than 8 μA_{rms}. This all-SiC neural probe realizes nearly monolithic integration of device components to provide a likely neurocompatible INI that should mitigate long-term reliability issues associated with chronic implantation.

Keywords: neural interface; neural probe; neural implant; microelectrode array; MEA; SiC; 3C-SiC; doped SiC; n-type; p-type; amorphous SiC; epitaxial growth; electrochemical characterization

1. Introduction

Implantable neural interfaces offer a method for external electronic devices to be connected to the central nervous system (CNS) in order to stimulate or record neurological signals, such as action potentials or multi-unit extracellular potentials, with the additional benefit of high spatial and temporal resolution. This interface forms a link for direct communication with the CNS through which the complex activities of neurons can be decoded to control active prosthetic devices or to stimulate one or more neural circuits to restore or enhance physiological functions [1].

Attempts to understand the electrophysiology of the nervous system started in the 17th century with stimulation of frog sciatic nerve [2]. Later in the 19th century, stainless steel wire electrode arrays were first implanted in the amygdala nuclei of monkeys and cats to investigate brain activity [3]. This was followed by the implantation of tungsten microelectrodes in the visual cortex of cats to investigate the behavior of individual cortical cells [4]. Study of the visual cortex, which requires a denser array of electrodes, drove a transition from metal wire electrode arrays to silicon-based three-dimensional microelectrode arrays (MEAs), such as the Utah array, which was introduced in the late 1980s [5,6]. This design minimized the electrode area and, as a result, allowed for higher spatial resolution during recording and stimulation of small populations of neurons, as well as utilized a reliable and repeatable manufacturing process. The high density of electrode sites, ability to individually address each electrode site, high-throughput fabrication, and compatibility with integrated circuit fabrication processes has made silicon an attractive material for high density, electrical neural interface applications.

A milestone in the development of silicon-based implantable intracortical neural interfaces (INIs) was the Michigan probe, introduced in 1970 [7], which employed multiple electrode sites on a single shank for chronic intracortical stimulation of, or recording from, single neurons [8]. Nevertheless, the occurrence of mechanical, material, and biological failures, both acute and chronic [9], has been a major factor in the questionability of silicon- and metallic-based micro-INIs for human utilization. Mechanical failure, in the form of lead or connector breakage, material degradation, or insulation delamination, and biological failures, such as bleeding, cell death, meningitis, gliosis, and fibrotic encapsulation and extrusion, have been reported elsewhere [10]. In one report, collected from an evaluation of 78 silicon-based intracortical MEAs chronically implanted in rhesus macaques, nearly half of the chronic failures happened within the first year [11]. The majority of those chronic failures (53%) were reported as biological failure caused by meningeal encapsulation and extrusion from the tissue. These results indicate the importance of a mechanically and chemically robust INI that offers better compatibility with the CNS to provide long-term recording and stimulation capabilities.

In recent years, researchers developing neural implants have turned their focus to flexible materials and designs to develop tissue-like INIs that address both mechanical and form factor compatibility. One implementation is an ultra-flexible, polymer-based probe in which a metal layer is sandwiched between two layers of SU-8 polymer [12]. Although this polymer-metal probe, and other similar designs [13,14], have shown a reduced immune response and were able to record action potentials and stimulate neurons, difficulty in the fabrication of these polymer-based devices, insertion of flexible polymer probes into the brain, and oxidation still remain fundamental issues [15]. Another method used to enhance the biocompatibility of neural probes are coatings that alter surface chemistry to provide hemostatic or immunomodulatory support [16]. In one example [17], a L1 protein coating was used to reduce microglial surface coverage. However, the surface coatings lose effectiveness over time leading to increased impedance and reduction in the recorded signals, and, in some cases, the mechanisms through which modulation of neurodegeneration and the corrosive behavior of encapsulating cells occurred was unclear [18].

For an INI to stimulate and record neural signals reliably over many years, both choice of material and their homogeneity must be carefully taken into consideration. Crystalline silicon carbide (SiC) is a semiconductor with a short bond length that gives it high physical resilience and chemical inertness. One of the important properties of SiC is that it displays polymorphism, which results in numerous single-crystal forms with the principal being hexagonal (i.e., 4H- and 6H-SiC) and cubic (i.e., 3C-SiC). SiC has been used in both the high-power electronics and MEMS industries [19,20]. It has also demonstrated a high degree of biological tolerance in vitro [21–24]. In addition, amorphous SiC (*a*-SiC), which provides excellent electrical insulation, has also shown good compatibility with neural cells [24–26] and has previously been used in the fabrication of several types of MEAs [27–32]. The properties of crystalline and amorphous SiC, and the results of previous studies, indicate that SiC can address the interrelated issues of INI biocompatibility and long-term reliability.

In our previous work, we reported the fabrication and characterization of nearly monolithic MEAs made from 4H-SiC, a hexagonal polytype of crystalline SiC, with *a*-SiC insulation [33]. However, the manufacture of these devices, as well as their release from the bulk SiC wafer, made these devices difficult to fabricate and costly. Here we report on the design and fabrication of a Michigan-style SiC neural probe on a silicon-on-insulator (SOI) wafer for ease of manufacture. The probe is composed of 3C-SiC, which was epitaxially grown on a SOI wafer. A heavily doped n-type (n^+) 3C-SiC film was grown on a moderately doped p-type SiC layer, forming a p-n junction. The n^+ layer was used to form the traces and electrode sites, eliminating the need for metallic conductive traces and metallic electrode sites that are in direct contact with the CNS tissue. The p-n junction structure provides substrate isolation between the conductive traces. A thin film of *a*-SiC was deposited via plasma-enhanced chemical vapor deposition (PECVD) on the probe to provide insulation from the external environment. The oxide buried in the SOI wafer served as a sacrificial layer, allowing the SiC probe to be released from the wafer with a selective wet etch process. This new fabrication approach, based on an all-SiC probe design, eliminates residual stresses typically found in similar devices consisting of stacks of heterogeneous films. It is expected that this approach will enhance the long-term material stability of implantable neural probes in the CNS, therefore increasing device reliability over many years. However, now that the manufacture of the probes has been demonstrated, follow-on studies in laboratory animals is required to support this hypothesis and are in the planning stages.

2. Materials and Methods

The all-SiC neural probe was developed using variations of standard silicon semiconductor micromachining processes. This started with epitaxial growth of a 3C-SiC film on a SOI wafer [20], followed by patterning of the 3C-SiC epitaxial films via thin film contact photolithography techniques. This was followed by the subsequent etching of features using a deep-reactive ion etcher (DRIE), deposition of a conformal *a*-SiC film via PECVD, and a final probe definition etch through the buried oxide layer via a DRIE process. The final step was the release of the finished device from the substrate SOI wafer by wet etching the buried oxide layer. The thickness of the doped epitaxial films was measured using cross-section scanning electron microscopy (SEM) and the composition was verified through energy-dispersive X-ray spectroscopy (EDS). No S-peak was observed in the EDS spectrum, indicating that the device surface was free of chemical residue from the etch processes. A commercial connector (Nano Strip, Omnetics Connector Corporation, Minneapolis, MN, USA) was used to make the electrical connections to the probe. Planar single-ended test microelectrodes were fabricated from the same epiwafer material as the implants for ease of electrical testing. Cyclic voltammetry (CV) and electrochemical impedance spectroscopy (EIS) in a phosphate buffered saline (PBS) solution, as well as p-n junction isolation and leakage current tests, were conducted on the test microelectrodes to electrically characterize the fabricated probes.

2.1. Epitaxial Growth of 3C-SiC on SOI

A 100 mm diameter SOI ((100) Si-oriented) wafer, with an ~26 μm silicon film on top of an ~2 μm buried thermal oxide layer, was used for fabrication of the all-SiC neural probes reported here. The growth process started with epitaxial growth of an ~8 μm p-type 3C-SiC film on the SOI wafer, followed by an ~2 μm heavily n-type (n^+) film, using a hot-wall reactor (LPE Epitaxial Technology, Baranzate, Italy) [34]. Heavy doping of semiconductors results in semi-metallic performance, which is the case for 3C-SiC. For this to be achieved, a n^+ doping density of ~10^{19} dopants/cm^3 is required. Hydrogen (H_2) was used as a carrier gas [19], ethylene (C_2H_4) as the carbon precursor, and trichlorosilane ($SiHCl_3$) as the silicon precursor gas. The epitaxial growth temperature was set to ~1370 °C with a process pressure of ~75 Torr. The C:Si ratio was kept between 0.8 and 1.2 throughout the epitaxial growth process. Aluminum and nitrogen were the p and n dopants, respectively, and were introduced during the epitaxial growth process [35,36]. The doping level of the top n^+ 3C-SiC film was measured with a LEI 2017b Mercury (Hg) Probe (Lehighton Electronics, Inc., Lehighton, PA, USA) [33].

2.2. Fabrication of All-SiC Neural Probe

The fabrication process sequence is shown in Figure 1. First, the epiwafer with the SiC films was cleaned in a solvent and then a RCA bath. It was then dipped in hydrofluoric acid (HF, 49%, J. T. Baker, Inc., Phillipsburg, NJ, USA) diluted in water (50:1) to remove any oxide that may have formed on top of the epitaxial 3C-SiC layer, followed by a DI water rinse and N_2 dry. Next, the wafer surface was functionalized with HMDS (Hexamethyldisilazane; Microchemicals GmBH, Ulm, Germany) and a 15–18 μm layer of AZ 12XT-20PL positive photoresist (Microchemicals GmBH) was spun on top at 1500 rpm. After a soft-bake at 110 °C for 3 min, the photoresist was patterned by UV exposure (110 mJ/cm^2) with a Quintel Mask Aligner and then baked at 90 °C for 1 min. The wafers were re-hydrated at ambient condition for 2 h and then developed with AZ300 developer (Microchemicals GmBH). The patterned photoresist was thick enough to allow for a ~3 μm deep etch of the epitaxial film using an Adixen AMS 100 DRIE. This process used oxygen (O_2) at 10 sccm and sulfur hexafluoride (SF_6) at 90 sccm. The pressure inside the chamber was set to 5.7 mTorr and the sample holder temperature was set to −20 °C. The sample holder power was kept at 550 W, while the source RF power was 1800 W. This process formed the traces and electrode sites on the probes.

A ~250 nm layer of *a*-SiC was deposited on the sample using PECVD (Unaxis 790, PlasmaTherm, Saint Petersburg, FL, USA). Methane (CH_4) and silane (SiH_4, 5% in He) were used as reactive gases to produce the *a*-SiC with flow rates of 200 sccm and 300 sccm, respectively. Helium (He), with a flow rate of 700 sccm, was used as the carrier gas. The RF power was set to 200 W, substrate temperature to 300 °C, and pressure to 1100 mTorr [37,38]. Following photoresist patterning using AZ 15nXT-450 CPS negative photoresist (Microchemicals GmbH), a reactive ion etch (RIE; PlasmaTherm) was run for 210 s to open windows in the *a*-SiC film for the electrode sites. Tetrafluoromethane (CF_4) and O_2, at 37 sccm and 13 sccm respectively, were used as the process gases. The power was set to 200 W and the chamber pressure to 50 mTorr. In order to package the probes for electrical testing, metal bonding pads were formed on one end of the traces (for the implants this is located on a tab that would reside outside the skull of the animal during in vivo testing). A 20 nm titanium (Ti) film, followed by a 200 nm gold (Au) film, was deposited without breaking vacuum in an electron beam evaporator and patterned using a lift-off process. Thermal annealing was performed to create an ohmic contact at the interface between the semiconductor and metal in a rapid thermal processor at 650 °C for 10 min [39]. This process sequence formed the contact pads for the commercial connector, which was used to connect the electrodes to external electronics.

The last step of the fabrication process was probe release. The same DRIE etch recipe that was used for formation of the traces was employed in an etch-through process to define the probes, except that the duration was increased to 15 minutes in order to ensure complete through etch of the 3C-SiC epitaxial films and top silicon layer. A scrap piece of the epiwafer was cleaved and cross-section SEM was used to determine the 3C-SiC epilayer and Si device layer thickness so that proper etch depth and mask thickness were selected. After removing the photoresist, the etch depth was measured using a contact profilometer (Dektak 150, Veeco, Plainview, NY, USA). The probes were released via wet etch of the sacrificial oxide layer with HF (49%). Then they were carefully removed from the HF solution, rinsed with DI water, and dried with N_2. To remove the backside silicon, the probes were adhered upside-down to a Si handle wafer with ~1 μm thermally grown oxide on top using a thin photoresist layer and placed in the DRIE. The residual Si was removed using the same DRIE recipe used for the definition of the electrodes and traces.

2.3. P-N Junction Isolation and Leakage Evaluation

Since p-n junctions are formed between the n^+ and p epitaxial films, back-to-back diodes are present between adjacent traces, which provides isolation. This isolation was evaluated by measuring the forward and reverse blocking voltages of test structures consisting of p-n diodes and n-p-n junctions formed between adjacent traces that were built on the 3C-SiC wafer. A Keithley 2400 SourceMeter (Tektronix, Inc., Beaverton, OR, USA) was used to generate current-voltage (I-V) plots for adjacent traces to observe these voltages. The voltage was increased from −10 V to +10 V at a rate of 0.1 V/s for the diodes and n-p-n junctions, and the observed currents recorded. The forward voltage was estimated using a semi-logarithmic current scale I-V plot [40]. The breakdown voltage occurs when the current rapidly increases during application of negative voltage. The root mean square (rms) of the current amplitude between breakdown and forward potentials for the diodes was defined as reverse leakage current [33]. The threshold current for defining the breakdown voltages was 10 μA.

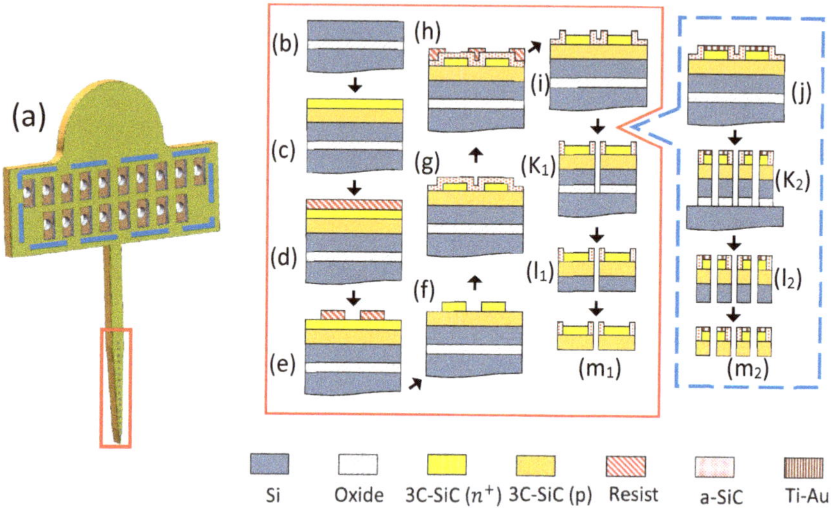

Figure 1. The all-SiC fabrication process flow. (**a**) A rendering of the Michigan-style 3C-SiC probe. The process flow inside the red rectangle shows the cross-section at the electrode sites while the blue rectangle provides the cross-section at the contact pads on the tab. (**b**) Starting SOI wafer, (**c**) ~8 μm of p-type 3C-SiC was grown on top, followed by ~2 μm of heavily n-type (n^+) 3C-SiC. (**d**) The wafer was coated with photoresist and (**e**) patterned via photolithography. (**f**) DRIE process was used to form the conductive n^+ mesas and (**g**) a thin *a*-SiC insulating layer was deposited on top via PECVD. (**h**) Photoresist was then patterned with photolithography and (**i**) the *a*-SiC was etched to form windows for the electrode sites using a RIE process. (**j**) After the *a*-SiC windows were opened, a layer of titanium, followed by gold, was deposited on the contact pads and thermally annealed. A deep DRIE etch through both epi layers and the oxide was performed to (k_1) define the probes and (k_2) form through-holes in the contact pads. (l_1, l_2) The oxide layer was etched in HF (49%) to release the probes. (m_1, m_2) Back-thinning via DRIE was performed to remove the residual silicon from the SOI device layer.

2.4. Electrochemical Characterization of All-SiC Probes

Electrochemical characterization of the 3C-SiC electrodes was performed via CV and EIS evaluation. A three-electrode setup was used with a potentiostat (VersaSTAT 4, AMETEK, Inc., Berwyn, PA, USA) to adjust the voltage between the working and counter electrodes in the presence of a reference electrode. CV provided information on the charge transfer properties of the electrode-electrolyte interface and on the presence of electrochemical reactions and their reversibility. Potential limits of −600 mV and +800 mV, which is the electrochemical window for platinum (Pt), were chosen for CV because this allowed for direct comparison of our measurements with previously published results [27,32,41,42]. EIS provided complex impedance measurements (both magnitude and phase) at frequencies of interest, which were used to evaluate the performance of the n^+ 3C-SiC conductor traces and electrodes.

Planar test microelectrodes fabricated alongside the neural probes on the same wafer were used for CV and EIS measurements [33,37]. The measurements were performed at room temperature in PBS with a pH of 7.40 ± 0.01, which was adjusted with hydrochloric acid (HCl). The PBS was composed of 137 mMol NaCl, 2.7 mMol KCl, and 10 mMol Na_2HPO_4. The gas levels in the PBS were ambient and no bubbling was done. The counter electrode was Pt and the reference electrode was Ag|AgCl. EIS measurements were performed from 0.1 Hz to 1 MHz with a rms voltage of 10 mV. The current was recorded 12 times per decade and three repetitions were averaged. CV measurement was initiated from open circuit potential, swept to −600 mV, and increased to +800 mV at a rate of 50 mV/s. This cycle was repeated three times and results were averaged. Charge values were calculated from the CV I-V curve via numerical integration with the trapezoidal method, trapz, in MATLAB (MathWorks, Natick, MA, USA).

3. Results

3.1. Epitaxial 3C-SiC Films

A cross-sectional SEM view of the wafer, which allows for accurate estimation of film thickness (n^+-, p-SiC, Si device film, and buried oxide), is shown in Figure 2a. This figure highlights various layers and the approximate thickness of each layer on the wafer used for the fabrication. The two epitaxial 3C-SiC films were measured, and their combined thickness determined to be ~10 µm. The SOI Si device layer (~26 µm), as well as the thin (~2 µm) buried oxide layer are also visible in this figure. The epitaxial n^+ film in the center of the wafer was quite rough with a mean surface roughness of ~244 nm and smoother near the wafer edge with a mean surface roughness of ~21 nm. Figure 2b shows surface morphology of the smooth n^+ layer, which was taken using a DI AFM (Dimension 3100). Although rough in the wafer center (Figure 2c), the surface roughness was low enough for thick layers of photoresist to properly cover the entire surface for the subsequent fabrication steps. However, this roughness would be expected to impact device electrical performance, particularly p-n diode leakage current.

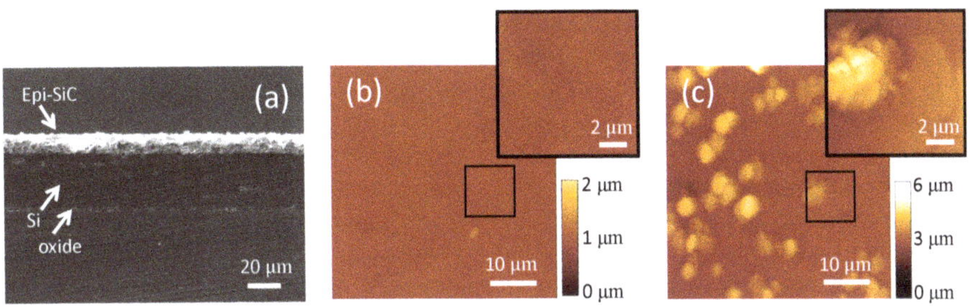

Figure 2. Analysis of epitaxial SiC results. (**a**) Cross-section SEM micrograph of the 3C-SiC epi films on SOI. (**b**) AFM image (tapping mode) of the 3C-SiC epiwafer specular region on the wafer periphery that shows typical 3C-SiC surface morphology (mean roughness of ~21 nm). (**c**) AFM image (tapping mode) of the rough surface of the same epiwafer (center) (mean roughness of ~244 nm). The devices were fabricated from the center of the wafer.

3.2. Fabricated All-SiC Neural Probe

Epitaxial growth of single crystalline 3C-SiC with different types of doping enables realization of a nearly monolithic probe from homogeneous SiC material. The all-SiC probe is a Michigan-style, planar neural probe with 16 electrodes for recording and stimulating neurons. The connector tab

was removed later via back-thinning using DRIE. This thick silicon layer can cause residual stress, due to mismatches in the coefficients of thermal expansion and lattice parameters [45] at the interface between the SiC films and silicon, resulting in bowing or bending of the probes. The SOI wafer used in this work had an ~20 um thick top silicon layer and this may have been the cause of the bowing of the shank and some warping in the connector tab. The shank should be straight for a proper insertion trajectory into the neural tissue. Also, in order to maximize contact at the connector interface, the tab containing the contact pads should be as planar as possible. Using a SOI wafer with a thin silicon device layer may resolve this deformation problem and will be used in future all-SiC devices.

Epitaxial 3C-SiC thin films are ceramic-like materials with, relative to neural tissue, a high elastic modulus, measured to be 424 ± 44 GPa using microsample tensile testing [46] and 433 ± 50 GPa using nanoindentation [47]. Defects can reduce the value the Young's modulus of 3C-SiC [48] and doping may affect this value as well [49]. There is a trend towards utilizing softer materials, such as polymers, for implantable neural interfaces due to their potential to improve the interaction with neural tissue [50–52]. By decreasing the Young's modulus of the neural probe closer to the values of neural tissues, the harmful shear and normal stress applied from the shank to the tissue should decrease. However, it is really device stiffness, which includes cross-sectional area, rather than just device modulus, that seems to matter the most [53]. Additionally, use of these softer materials introduce challenges with fabrication processes, scaling to higher channel-count systems, particularly with respect to interconnects, and can lead to insertion difficulties. Once implanted, these materials face challenges with material stability and device reliability [54]. The hard, chemically inert nature, and ease of micromachining with traditional silicon processes means SiC neural probes may suffer less from these limitations. Clearly, long-term in vivo studies in an animal model are needed to assess the performance of the all-SiC INI and are planned.

It has been demonstrated that once the implanted structure size is reduced to subcellular scale, i.e., less than ~10 μm, the foreign body response and associated neuron death is greatly reduced in a rat model [55,56]. With traditional silicon probes, reducing size increases the occurrence of probe fracture at high stress regions [57]. SiC is a much more robust material, with a reduced tendency to fracture at these desired smaller sizes, while maintaining the mechanical strength needed for proper penetration of neural tissue [24,27]. Therefore, SiC is an excellent material for developing a high electrode density neural interface, allowing for further reduction in size while greatly minimizing risk of fracture.

The heterogeneous composition of implanted neural interfaces that utilize metallic materials as electrode sites or conductive traces may increase the risk of delamination in chronic implantation, specifically, at regions under higher stress [57]. Delamination usually occurs at the interface between metal and semiconductor materials due to residual stress in the thin films. A homogeneous material composition can eliminate this residual stress, reducing the risk of delamination at the interfaces between different materials in the probe by eliminating them.

The 3C-SiC is a wide-band-gap semiconductor with a high band energy of ~2.2 eV. This results in a higher turn-on voltage at the junction between n^+- and p-type SiC. This higher turn-on voltage provides a wider voltage range to stimulate neurons while isolating individual channels via n-p-n junctions supporting simultaneous multichannel microstimulation and recording, as might be necessary for implementing a closed-loop system. The turn-on voltage for p-n junctions built from Si is ~0.7 V, which is low compared to ~1.4 V for SiC, and limits proper isolation via a n-p-n junction configuration. However, the higher leakage current in our all-SiC films may negatively affect the final device's functionality. Surface roughness is known to be associated with the density of crystal defects, thus a higher defect density may cause higher leakage current [43]. It is believed that the high surface roughness in this work, an indication of poor crystallinity, in conjunction with a high number of defects, may be the cause of the observed high leakage current. For reference, in our 4H-SiC devices with specular surface morphology, the leakage current was nA versus μA reported here [33]. A lower surface roughness via an optimized epitaxial growth process would be expected to improve both the mechanical properties and the leakage current [58].

The EIS results revealed that the doped, semi-metallic 3C-SiC conductors have impedance values approaching those of metals commonly used in implantable microelectrodes, such as gold, platinum, or tungsten, as well as highly doped polysilicon [59,60]. The average impedance for a surface area of 491 μm^2 was approximately 75% lower (165 kΩ vs. 675 kΩ at 1kHz) than previously reported for our 4H-SiC electrodes [33].

Both the charge balanced CV cycles and the negative phase angles from EIS measurements support a dominant capacitive charge transfer mechanism for 1 kHz and higher frequencies at the electrode-electrolyte interface, but faradaic currents may be present at lower frequencies. This differs from capacitive electrode materials like titanium nitride (TiN) [61], which has a phase closer to 90° at lower frequencies [62]. Compared to values previously reported for 4H-SiC, the charge values calculated from CV measurements reported here were approximately two orders of magnitude higher, with the average charge storage capacity (anodic: 15 mC/cm^2 vs. 0.41 mC/cm^2; cathodic: 15 mC/cm^2 vs. 0.19 mC/cm^2) and an average charge per phase (anodic: 75 nC vs. 2.0 nC; cathodic: 75 nC vs. 1.0 nC) using a Pt electrochemical window (−600 mV to +800 mV). It is possible that the greater surface roughness accounts for this large difference in electrochemical properties. It is also possible that there were more faradaic reactions at lower frequencies leading to more oxidation and reduction at the surface, which may be linked to defect sites in the SiC.

Current neural probe technology built from materials like silicon suffer from long-term reliability issues that reduces their lifetime considerably, resulting in loss of recording and microstimulation function when chronically implanted. This limits their use in medical applications for humans. Device-based modalities could become a more common alternative to pharmaceuticals for treatment of neurological trauma or disease if the issue of long-term reliability in implantable neural interfaces is properly addressed. After further refinement of the design and optimization of the material processing, the performance of the all-SiC neural probe will be evaluated with chronic *in vivo* experiments in rodent models to investigate its long-term safety and effectiveness in neural tissue. There is accumulating evidence [25–27,29,30,32,63] that SiC could be an appropriate material for the greatly needed implantable neural interface that functions for the lifetime of the recipient.

5. Conclusions

The fabrication and initial electrical characterization of an all-SiC neural probe is presented. The SiC neural probe was fabricated from p− and n$^+$-type 3C-SiC epilayers grown on SOI wafers. First, a moderately p-type 3C-SiC film was grown on a SOI wafer, followed by a layer of n$^+$-type 3C-SiC. The surface morphology of the top n$^+$ epilayer was measured. Neural probes with sixteen traces, electrode sites, and other test structures were patterned on the 3C-SiC epilayers via MEMS microfabrication processes. Metallic traces were absent from the shank of the probe, and instead a semi-metallic n$^+$ layer was formed into traces and electrode sites. A thin layer of *a*-SiC film was deposited on top of the epilayers to serve as an insulator. The probes were harvested using dissolution of the buried oxide layer in the SOI handle wafer to provide ease of manufacture. The backside silicon layer remaining after release of the probes was removed via back-thinning in a DRIE. Adjacent traces were electrically isolated through a n-p-n junction. After completion of device fabrication, the performance of the n-p-n junctions was evaluated through current-voltage measurements and the turn-on voltage was determined to be ~1.4 V. Electrical measurements showed satisfactory p-n junction performance, but leakage current needs to be improved via higher quality 3C-SiC epitaxial films. In addition, initial electrochemical characterization work with 491 μm^2 surface area test microelectrodes demonstrated good impedance, charge storage capacity, and charge per phase values. These results support the feasibility of neural stimulation and recording with the fabricated all-SiC neural probe. However, further studies are necessary to demonstrate the acute recording and stimulation capability and chronic stability of the fabricated SiC neural probes, and, consequently, in vitro accelerated aging and in vivo studies in a rodent model are planned and will be reported in the future.

Author Contributions: M.B. conducted all device fabrication steps except for *a*-SiC deposition. He also performed dry electrical testing of the p-n and n-p-n structures and co-wrote the manuscript. J.T.B. was responsible for electrochemical characterization, performed all EIS and CV measurements, and co-wrote the manuscript. C.A.K. carried out the *a*-SiC deposition and provided XPS analysis of the films on blank companion samples. C.L.F. and S.E.S. developed the all-SiC INI concept, device design, and provided technical consulting to the fabrication team. E.J.B. initiated the 3C-SiC INI device fabrication at USF prior to graduating with his doctorate in 2018 and provided technical consultation to the fabrication team on this work. F.L.V. is an expert on 3C-SiC epitaxial growth and provided the epitaxial wafers used in this work along with technical consultation on the device characterization performed. A.T. is an expert on electrochemistry and provided technical consultation on the EIS and CV measurements and analysis. All authors reviewed and edited the final draft of the manuscript.

Funding: Funding at the University of South Florida was provided via teaching assistantships (M. Beygi and J. Bentley) and from Dr. Saddow's overhead funds to purchase consumable items.

Acknowledgments: The *a*-SiC films were grown in the NIMET at the University of Florida under the supervision of Jack Judy and his support is gratefully acknowledged. The assistance of Chenyin Feng with SEM image acquisition and Richard Everly with microfabrication processes in the University of South Florida Nanotechnology Research and Education Center (NREC) is appreciated.

Conflicts of Interest: The authors declare no conflict of interest.

References

1. Öztürk, S.; Devecioğlu, İ.; Beygi, M.; Atasoy, A.; Mutlu, Ş.; Özkan, M.; Güçlü, B. Real-Time Performance of a Tactile Neuroprosthesis on Awake Behaving Rats. *IEEE Trans. Neural Syst. Rehabil. Eng.* **2019**, *27*, 1053–1062. [CrossRef] [PubMed]
2. Verkhratsky, A.; Krishtal, O.A.; Petersen, O.H. From Galvani to patch clamp: the development of electrophysiology. *Pflügers Arch.* **2006**, *453*, 233–247. [CrossRef] [PubMed]
3. Delgado, J.M. Permanent implantation of multilead electrodes in the brain. *Yale J. Biol. Med.* **1952**, *24*, 351. [PubMed]
4. Hubel, D.H.; Wiesel, T.N. Receptive fields, binocular interaction and functional architecture in the cat's visual cortex. *J. Physiol.* **1962**, *160*, 106–154. [CrossRef] [PubMed]
5. Campbell, P.K.; Jones, K.E.; Huber, R.J.; Horch, K.W.; Normann, R.A. A silicon-based, three-dimensional neural interface: manufacturing processes for an intracortical electrode array. *IEEE Trans. Biomed. Eng.* **1991**, *38*, 758–768. [CrossRef] [PubMed]
6. Normann, R.A.; Campbell, P.K.; Li, W.P. Silicon based microstructures suitable for intracortical electrical stimulation (visual prosthesis application). In Proceedings of the Annual International Conference of the IEEE Engineering in Medicine and Biology Society, New Orleans, LA, USA, 4–7 November 1988; Volume 712, pp. 714–715.
7. Wise, K.D.; Angell, J.B.; Starr, A. An Integrated-Circuit Approach to Extracellular Microelectrodes. *IEEE Trans. Biomed. Eng.* **1970**, *BME-17*, 238–247. [CrossRef]
8. Vetter, R.J.; Williams, J.C.; Hetke, J.F.; Nunamaker, E.A.; Kipke, D.R. Chronic neural recording using silicon-substrate microelectrode arrays implanted in cerebral cortex. *IEEE Trans. Biomed. Eng.* **2004**, *51*, 896–904. [CrossRef]
9. Gunasekera, B.; Saxena, T.; Bellamkonda, R.; Karumbaiah, L. Intracortical Recording Interfaces: Current Challenges to Chronic Recording Function. *ACS Chem. Neurosci.* **2015**, *6*, 68–83. [CrossRef]
10. Jorfi, M.; Skousen, J.L.; Weder, C.; Capadona, J.R. Progress towards biocompatible intracortical microelectrodes for neural interfacing applications. *J. Neural Eng.* **2015**, *12*, 011001. [CrossRef]
11. Barrese, J.C.; Rao, N.; Paroo, K.; Triebwasser, C.; Vargas-Irwin, C.; Franquemont, L.; Donoghue, J.P. Failure mode analysis of silicon-based intracortical microelectrode arrays in non-human primates. *J. Neural Eng.* **2013**, *10*, 066014. [CrossRef]
12. Xie, C.; Liu, J.; Fu, T.-M.; Dai, X.; Zhou, W.; Lieber, C.M. Three-dimensional macroporous nanoelectronic networks as minimally invasive brain probes. *Nat. Mater.* **2015**, *14*, 1286. [CrossRef] [PubMed]
13. Thukral, A.; Ershad, F.; Enan, N.; Rao, Z.; Yu, C. Soft Ultrathin Silicon Electronics for Soft Neural Interfaces: A Review of Recent Advances of Soft Neural Interfaces Based on Ultrathin Silicon. *IEEE Nanotechnol. Mag.* **2018**, *12*, 21–34. [CrossRef]
14. Scholten, K.; Meng, E. Materials for microfabricated implantable devices: a review. *Lab. Chip* **2015**, *15*, 4256–4272. [CrossRef] [PubMed]

15. Hong, G.; Lieber, C.M. Novel electrode technologies for neural recordings. *Nat. Rev. Neurosci* **2019**. [CrossRef] [PubMed]
16. Oakes, R.S.; Polei, M.D.; Skousen, J.L.; Tresco, P.A. An astrocyte derived extracellular matrix coating reduces astrogliosis surrounding chronically implanted microelectrode arrays in rat cortex. *Biomaterials* **2018**, *154*, 1–11. [CrossRef] [PubMed]
17. Eles, J.R.; Vazquez, A.L.; Snyder, N.R.; Lagenaur, C.; Murphy, M.C.; Kozai, T.D.; Cui, X.T. Neuroadhesive L1 coating attenuates acute microglial attachment to neural electrodes as revealed by live two-photon microscopy. *Biomaterials* **2017**, *113*, 279–292. [CrossRef] [PubMed]
18. Wellman, S.M.; Eles, J.R.; Ludwig, K.A.; Seymour, J.P.; Michelson, N.J.; McFadden, W.E.; Vazquez, A.L.; Kozai, T.D.Y. A Materials Roadmap to Functional Neural Interface Design. *Adv. Funct Mater.* **2018**, 28. [CrossRef] [PubMed]
19. Wijesundara, M.; Azevedo, R. *Silicon Carbide Microsystems for Harsh Environments*; Springer Science & Business Media: Berlin/Heidelberg, Germany, 2011; Volume 22.
20. Sarro, P.M. Silicon carbide as a new MEMS technology. *Sens. Actuators A Phys.* **2000**, *82*, 210–218. [CrossRef]
21. Coletti, C.; Jaroszeski, M.; Pallaoro, A.; Hoff, A.; Iannotta, S.; Saddow, S. Biocompatibility and wettability of crystalline SiC and Si surfaces. In Proceedings of the 2007 29th Annual International Conference of the IEEE Engineering in Medicine and Biology Society, Lyon, France, 22–26 August 2007; pp. 5849–5852.
22. Saddow, S.E.; Frewin, C.; Reyes, M.; Register, J.; Nezafati, M.; Thomas, S. 3C-SiC on Si: A Biocompatible Material for Advanced Bioelectronic Devices. *ECS Trans.* **2014**, *61*, 101–111. [CrossRef]
23. Kotzar, G.; Freas, M.; Abel, P.; Fleischman, A.; Roy, S.; Zorman, C.; Moran, J.M.; Melzak, J. Evaluation of MEMS materials of construction for implantable medical devices. *Biomaterials* **2002**, *23*, 2737–2750. [CrossRef]
24. Frewin, C.L.; Coletti, C.; Register, J.J.; Nezafati, M.; Thomas, S.; Saddow, S.E. Silicon Carbide Materials for Biomedical Applications. In *Carbon for Sensing Devices*; Elsevier: Amsterdam, The Netherlands, 2015; pp. 153–207.
25. Knaack, G.L.; McHail, D.G.; Borda, G.; Koo, B.; Peixoto, N.; Cogan, S.F.; Dumas, T.C.; Pancrazio, J.J. In vivo Characterization of Amorphous Silicon Carbide as a Biomaterial for Chronic Neural Interfaces. *Front. Neurosci.* **2016**, *10*, 301. [CrossRef] [PubMed]
26. Knaack, G.L.; Charkhkar, H.; Cogan, S.F.; Pancrazio, J.J. Amorphous silicon carbide for neural interface applications. In *Silicon Carbide Biotechnology*; Elsevier: Amsterdam, The Netherlands, 2016; pp. 249–260.
27. Deku, F.; Frewin, C.L.; Stiller, A.; Cohen, Y.; Aqeel, S.; Joshi-Imre, A.; Black, B.; Gardner, T.J.; Pancrazio, J.J.; Cogan, S.F. Amorphous Silicon Carbide Platform for Next Generation Penetrating Neural Interface Designs. *Micromachines* **2018**, *9*, 480. [CrossRef] [PubMed]
28. Diaz-Botia, C.; Luna, L.; Chamanzar, M.; Carraro, C.; Sabes, P.; Maboudian, R.; Maharbiz, M. Fabrication of all-silicon carbide neural interfaces. In Proceedings of the 2017 8th International IEEE/EMBS Conference on Neural Engineering (NER), Shanghai, China, 25–28 May 2017; pp. 170–173.
29. Deku, F.; Cohen, Y.; Joshi-Imre, A.; Kanneganti, A.; Gardner, T.J.; Cogan, S.F. Amorphous silicon carbide ultramicroelectrode arrays for neural stimulation and recording. *J. Neural Eng.* **2018**, *15*, 016007. [CrossRef] [PubMed]
30. Deku, F.; Ghazavi, A.; Cogan, S.F. Neural interfaces based on amorphous silicon carbide ultramicroelectrode arrays. *Bioelectron. Med.* **2018**, *1*, 185–200. [CrossRef]
31. Frewin, C.; Locke, C.; Mariusso, L.; Weeber, E.; Saddow, S. Silicon carbide neural implants: In vivo neural tissue reaction. In Proceedings of the 2013 6th International IEEE/EMBS Conference on Neural Engineering (NER), San Diego, CA, USA, 6–8 November 2013; pp. 661–664.
32. Diaz-Botia, C.A.; Luna, L.E.; Neely, R.M.; Chamanzar, M.; Carraro, C.; Carmena, J.M.; Sabes, P.N.; Maboudian, R.; Maharbiz, M.M. A silicon carbide array for electrocorticography and peripheral nerve recording. *J. Neural Eng.* **2017**, *14*, 056006. [CrossRef] [PubMed]
33. Bernardin, E.K.; Frewin, C.L.; Everly, R.; Ul Hassan, J.; Saddow, S.E. Demonstration of a Robust All-Silicon-Carbide Intracortical Neural Interface. *Micromachines* **2018**, *9*, 412. [CrossRef]
34. Zielinski, M.; Arvinte, R.; Chassagne, T.; Michon, A.; Portail, M.; Kwasnicki, P.; Konczewicz, L.; Contreras, S.; Juillaguet, S.; Peyre, H. p-Type Doping of 4H- and 3C-SiC Epitaxial Layers with Aluminum. *Mater. Sci. Forum.* **2016**, *858*, 137–142. [CrossRef]

35. La Via, F.; Severino, A.; Anzalone, R.; Bongiorno, C.; Litrico, G.; Mauceri, M.; Schoeler, M.; Schuh, P.; Wellmann, P. From thin film to bulk 3C-SiC growth: Understanding the mechanism of defects reduction. *Mater. Sci. Semicond. Process.* **2018**, *78*, 57–68. [CrossRef]
36. Zimbone, M.; Mauceri, M.; Litrico, G.; Barbagiovanni, E.G.; Bongiorno, C.; La Via, F. Protrusions reduction in 3C-SiC thin film on Si. *J. Cryst. Growth* **2018**, *498*, 248–257. [CrossRef]
37. Bernardin, E.; Frewin, C.L.; Dey, A.; Everly, R.; Hassan, J.U.; Janzén, E.; Pancrazio, J.; Saddow, S.E. Development of an all-SiC neuronal interface device. *MRS Adv.* **2016**, *1*, 3679–3684. [CrossRef]
38. Saddow, S.; Frewin, C.; Nezafati, M.; Oliveros, A.; Afroz, S.; Register, J.; Reyes, M.; Thomas, S. 3C-SiC on Si: A bio-and hemo-compatible material for advanced nano-bio devices. In Proceedings of the 2014 IEEE 9th Nanotechnology Materials and Devices Conference (NMDC), Sicily, Italy, 12–15 October 2014; pp. 49–53.
39. Roy, S.; Jacob, C.; Basu, S. Ohmic contacts to 3C-SiC for Schottky diode gas sensors. *Solid-State Electron.* **2003**, *47*, 2035–2041. [CrossRef]
40. Avila, R.; Kopanski, J.; Fung, C. Behavior of ion-implanted junction diodes in 3C SiC. *J. Appl. Phys.* **1987**, *62*, 3469–3471. [CrossRef]
41. McCreery, D.B.; Agnew, W.F.; Yuen, T.G.H.; Bullara, L. Charge density and charge per phase as cofactors in neural injury induced by electrical stimulation. *IEEE Trans. Biomed. Eng.* **1990**, *37*, 996–1001. [CrossRef] [PubMed]
42. Rose, T.L.; Robblee, L.S. Electrical stimulation with Pt electrodes. VIII. Electrochemically safe charge injection limits with 0.2 ms pulses (neuronal application). *IEEE Trans. Biomed. Eng.* **1990**, *37*, 1118–1120. [CrossRef] [PubMed]
43. Kong, H.-S.; Glass, J.; Davis, R.F. Growth rate, surface morphology, and defect microstructures of β–SiC films chemically vapor deposited on 6H–SiC substrates. *J. Mater. Res.* **1989**, *4*, 204–214. [CrossRef]
44. Maboudian, R.; Carraro, C.; Senesky, D.G.; Roper, C.S. Advances in silicon carbide science and technology at the micro-and nanoscales. *J. Vac. Sci. Technol. A Vac. Surf. Film* **2013**, *31*, 050805. [CrossRef]
45. Davis, R.F.; Sitar, Z.; Williams, B.; Kong, H.; Kim, H.; Palmour, J.; Edmond, J.; Ryu, J.; Glass, J.; Carter, C., Jr. Critical evaluation of the status of the areas for future research regarding the wide band gap semiconductors diamond, gallium nitride and silicon carbide. *Mater. Sci. Eng. B* **1988**, *1*, 77–104. [CrossRef]
46. Jackson, K.M.; Dunning, J.; Zorman, C.A.; Mehregany, M.; Sharpe, W.N. Mechanical properties of epitaxial 3C silicon carbide thin films. *J. Microelectromech. Syst.* **2005**, *14*, 664–672. [CrossRef]
47. Reddy, J.D.; Volinsky, A.A.; Frewin, C.L.; Locke, C.; Saddow, S.E. Mechanical Properties of 3C-SiC Films for MEMS Applications. *MRS Proc.* **2007**, 1049. [CrossRef]
48. Anzalone, R.; Camarda, M.; Canino, A.; Piluso, N.; La Via, F.; D'Arrigo, G. Defect Influence on Heteroepitaxial 3C-SiC Young's Modulus. *Electrochem. Solid-State Lett.* **2011**, *14*, H161–H162. [CrossRef]
49. Su, C.M.; Wuttig, M.; Fekade, A.; Spencer, M. Elastic and anelastic properties of chemical vapor deposited epitaxial 3C-SiC. *J. Appl. Phys.* **1995**, *77*, 5611–5615. [CrossRef]
50. Green, R.A.; Lovell, N.H.; Wallace, G.G.; Poole-Warren, L.A. Conducting polymers for neural interfaces: Challenges in developing an effective long-term implant. *Biomaterials* **2008**, *29*, 3393–3399. [CrossRef] [PubMed]
51. Ware, T.; Simon, D.; Rennaker, R.L.; Voit, W. Smart Polymers for Neural Interfaces. *Polym. Rev.* **2013**, *53*, 108–129. [CrossRef]
52. Green, R.; Abidian, M.R. Conducting Polymers for Neural Prosthetic and Neural Interface Applications. *Adv. Mater.* **2015**, *27*, 7620–7637. [CrossRef] [PubMed]
53. Stiller, M.A.; Black, J.B.; Kung, C.; Ashok, A.; Cogan, F.S.; Varner, D.V.; Pancrazio, J.J. A Meta-Analysis of Intracortical Device Stiffness and Its Correlation with Histological Outcomes. *Micromachines* **2018**, *9*, 443. [CrossRef]
54. Woeppel, K.; Yang, Q.; Cui, X.T. Recent advances in neural electrode–tissue interfaces. *Curr. Opin. Biomed. Eng.* **2017**, *4*, 21–31. [CrossRef] [PubMed]
55. Kozai, T.D.Y.; Langhals, N.B.; Patel, P.R.; Deng, X.; Zhang, H.; Smith, K.L.; Lahann, J.; Kotov, N.A.; Kipke, D.R. Ultrasmall implantable composite microelectrodes with bioactive surfaces for chronic neural interfaces. *Nat. Mater.* **2012**, *11*, 1065. [CrossRef]
56. Pancrazio, J.J.; Deku, F.; Ghazavi, A.; Stiller, A.M.; Rihani, R.; Frewin, C.L.; Varner, V.D.; Gardner, T.J.; Cogan, S.F. Thinking Small: Progress on Microscale Neurostimulation Technology. *Neuromodul. Technol. Neural Interface* **2017**, *20*, 745–752. [CrossRef]

57. Kozai, T.D.; Catt, K.; Li, X.; Gugel, Z.V.; Olafsson, V.T.; Vazquez, A.L.; Cui, X.T. Mechanical failure modes of chronically implanted planar silicon-based neural probes for laminar recording. *Biomaterials* **2015**, *37*, 25–39. [CrossRef]
58. Morkoc, H.; Strite, S.; Gao, G.; Lin, M.; Sverdlov, B.; Burns, M. Large-band-gap SiC, III-V nitride, and II-VI ZnSe-based semiconductor device technologies. *J. Appl. Phys.* **1994**, *76*, 1363–1398. [CrossRef]
59. Saha, R.; Jackson, N.; Patel, C.; Muthuswamy, J. Highly Doped Polycrystalline Silicon Microelectrodes Reduce Noise in Neuronal Recordings In Vivo. *IEEE Trans. Neural Syst. Rehabil. Eng.* **2010**, *18*, 489–497. [CrossRef] [PubMed]
60. Howlader, M.M.R.; Alam, A.U.; Sharma, R.P.; Deen, M.J. Materials analyses and electrochemical impedance of implantable metal electrodes. *Phys. Chem. Chem. Phys.* **2015**, *17*, 10135–10145. [CrossRef] [PubMed]
61. Meijs, S.; Alcaide, M.; Sørensen, C.; McDonald, M.; Sørensen, S.; Rechendorff, K.; Gerhardt, A.; Nesladek, M.; Rijkhoff, N.J.; Pennisi, C.P. Biofouling resistance of boron-doped diamond neural stimulation electrodes is superior to titanium nitride electrodes in vivo. *J. Neural Eng.* **2016**, *13*, 056011. [CrossRef] [PubMed]
62. Weiland, J.D.; Anderson, D.J.; Humayun, M.S. In vitro electrical properties for iridium oxide versus titanium nitride stimulating electrodes. *IEEE Trans. Biomed. Eng.* **2002**, *49*, 1574–1579. [CrossRef] [PubMed]
63. Joshi-Imre, A.; Black, B.J.; Abbott, J.; Kanneganti, A.; Rihani, R.; Chakraborty, B.; Danda, V.R.; Maeng, J.; Sharma, R.; Rieth, L.; et al. Chronic recording and electrochemical performance of amorphous silicon carbide-coated Utah electrode arrays implanted in rat motor cortex. *J. Neural Eng.* **2019**. [CrossRef]

micromachines

MDPI

Article

Modeling and Experiment of the Critical Depth of Cut at the Ductile–Brittle Transition for a 4H-SiC Single Crystal

Peng Chai, Shujuan Li * and Yan Li *

School of Mechanical and Precision Instrument Engineering, Xi'an University of Technology, No. 5 South Jinhua Road, Xi'an 710048, Shaanxi, China; chaipeng@stu.xaut.edu.cn
* Correspondence: shujuanli@xaut.edu.cn (S.L.); jyxy-ly@xaut.edu.cn (Y.L.)

Received: 19 April 2019; Accepted: 5 June 2019; Published: 7 June 2019

Abstract: In this paper, a theoretical model of the critical depth of cut of nanoscratching on a 4H-SiC single crystal with a Berkovich indenter is proposed, and a series of scratch tests in a nanomechanical test system was performed. Through nanoindentation experimentation on fused quartz, the Berkovich indenter nose radius was indirectly confirmed using least squares. The range of critical depths of cut at the ductile–brittle transition was obtained by SEM observation, and the size of cracks was amplified with increasing scratching depth. The theoretical result of the critical depth of cut at the ductile–brittle transition for a 4H-SiC single crystal is 91.7 nm, which is close to the first obvious pop-in point of the relation curve between tangential force and lateral displacement. Repeated experimental results show good consistency and good agreement with other references.

Keywords: 4H-SiC; critical depth of cut; Berkovich indenter; cleavage strength; nanoscratching

1. Introduction

With excellent electronic characteristics such as a large band gap, high critical breakdown strength, high electronic saturation rate, high thermal conductivity, and high irradiation resistance, silicon carbide (SiC) has become an outstanding representative of the third generation of semiconductor materials and has been increasingly widely applied in a variety of fields, including computer, aviation, power, and nuclear energy development [1–4]. In addition, it is an attractive option to use SiC to produce space mirrors and large ground-based reflectors thanks to its remarkable advantages of large stiffness, small thermal deformation coefficient, and good thermal stability [5]. However, it is difficult to obtain SiC parts with high forming accuracy and ideal machined surface quality due to its characteristics of high hardness (with a Mohs hardness between 9.0 and 9.5) and brittleness [6,7].

The traditional cutting force models tend to ignore the effect of elastic recovery on the macroprocessing of ductile materials. However, numerous studies [8–10] have shown that the role of elastic recovery is significant in the micro–nano-machining of brittle materials. Wasmer et al. [11] proposed a typical scratch pattern for brittle materials using an increasing load along a scratch path that is divided into five stages: elastic regime, ductile regime, subsurface crack regime, surface crack regime, and micro-abrasive regime. Elastic and ductile deformations were observed before the ductile regime, and brittle fractures appeared and began to dominate the latter deformation with increasing cutting force. Therefore, there is a cutting depth, referred to as the critical depth of cut, where the ductile region transitions to the brittle region. Lawn et al. [12] established a microfracture model under the point indentation of brittle materials consisting of the following processes: (1) the sharp indenter induces a zone of elastic and ductile deformation around the contact point; (2) a median crack suddenly initiates below the contact point; (3) the median crack stably extends with increasing indenter load;

(4) the median crack begins to close during the initial unloading process; (5) lateral cracks begin to appear due to residual stress; and (6) lateral cracks continue to extend and cause chipping.

Methods such as grinding and chemical mechanical polishing are used in traditional processing and are characterized by low production efficiency, high production cost, and, particularly, surface damage caused by the contamination of the polishing slurry [13]. In 1951, researchers found that hard and brittle materials show the characteristics of ductile removal under certain processing conditions [14]. Since the 1990s, researchers have conducted many studies into removal mechanisms in silicon carbide ductile regime processing, such as ductile regime grinding [7,15], ductile regime laser-assisted processing [16], ductile regime diamond cutting [17], ductile regime diamond wiresaw [18], and ductile domain ultrasonic-assisted processing [19]. These studies indicated that in the course of ductile regime processing, the chip is removed by ductile deformation, causing no damage or cracks to the machined surface of the workpiece, and the surface processing quality can be maintained [20–22]. The critical depth of cut at the ductile–brittle transition, the maximum cutting depth where no cracks appeared on the surface or subsurface of the sample, is a fundamental parameter of all methods of ductile regime processing. A formula was obtained by Bifano according to Griffith's principle through a quasi-static scratch test of several kinds of common brittle materials [23]:

$$d_c = \alpha\left(\frac{E}{H}\right)\left(\frac{K_C}{H}\right)^2) \quad (1)$$

where α is a constant, E is the elasticity modulus, H is the hardness, and K_c is the fracture toughness. This formula was amended by later scholars; for example, Gaobo's study on the critical depth of cut of 6H-silicon carbide indicated that the experimental results were not in line with the calculation of Formula (1) and the amended constant α [24].

Based on the above, most research has focused on the critical depth of cut using Griffith's principle and experimental method, which is based on cracks in the ductile extension which would have an influence on the performance of devices. With the development of nanotechnology, particularly the scanning electron microscope (SEM) and nanomechanical testing technology, a number of powerful tools has been provided to investigate the properties of silicon carbide at the nanoscale. This paper proposes a method considering the elastic recovery of the workpiece surface in nanoscratching in order to obtain the critical depth of cut for SiC using scratching stress and cleavage strength.

2. Modeling

2.1. Modeling of the Indenter Structure

The indenter tip shape greatly affects nanoscratching results, and there is no ideal Berkovich indenter due to the limitations of processing conditions. In addition, the indenter continuously wears as it is working; therefore, the indenter tip shape is different for every test. The geometric shape and dimension parameters of a Berkovich indenter are shown in Figure 1. Assuming that a Berkovich indenter is a combination of a sphere and a triangular pyramid [25], the tip can be divided into three parts: the sphere (from Section 1 to the vertex of the nose in Figure 1a), the transition (from Section 1 to Section 2), and the pyramid (above Section 2). The following equations can be obtained according to the geometric relations:

$$d^* = \frac{R}{\sin\alpha} - R \quad (2)$$

$$d_2 = R(1 - \sin\alpha) \quad (3)$$

$$r_0{}^2 = R^2 - (R - d_1)^2 \quad (4)$$

$$\tan\theta = \frac{r_0}{d^* + d_1} \quad (5)$$

$$r = \sqrt{R^2 - (R + d*)^2 \sin^2\theta} \quad (6)$$

$$l = 2\sqrt{r^2 - \left(r - \frac{d - d_1}{\cos\theta}\right)^2} \tag{7}$$

where R is the indenter nose radius, d^* is the distance from the nose vertex to the top of the ideal indenter, α is the angle between the edge line and the centerline, θ is the angle between the edge plane and the centerline, d_1 is the distance from Section 1 to the nose vertex, d_2 is the distance from Section 2 to the nose vertex, r_0 is the radius of Section 1, r is the circular arc radius of the intersection between the sphere and the triangular pyramid, and l is the length of the intersection between the section which is normally aligned to the centerline and the edge plane in the transition part.

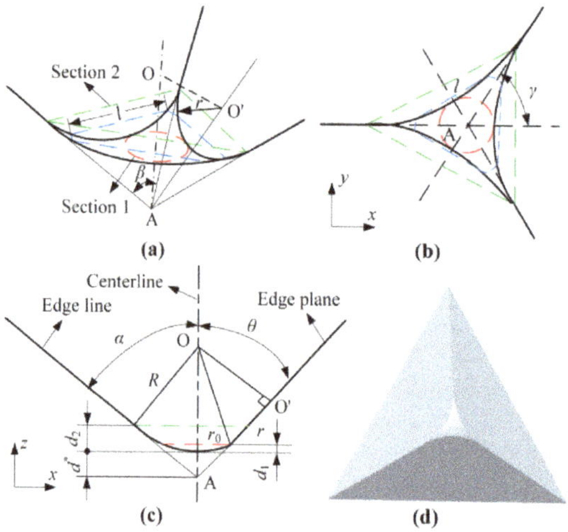

Figure 1. Dimension parameters and geometric shape of the Berkovich indenter: (**a**) model diagram, (**b**) top view, (**c**) side view, and (**d**) 3-D solid model.

The normal projected area of the different indenter heights can be calculated using Equation (8).

$$A_p = \begin{cases} \pi[R^2 - (R-d)^2] & (d \le d_1) \\ \pi \tan^2\alpha(d+d^*)^2 - \frac{\arcsin\frac{l}{2\tan\alpha(d+d^*)}}{60}\cdot\pi\tan^2\alpha(d+d^*)^2 + \frac{3l}{2}\tan\alpha(d+d^*) & (d_1 < d < d_2) \\ \frac{3\sqrt{3}}{4}\tan^2\alpha\left(d + \frac{R}{\sin\alpha} - R\right)^2. & (d \ge d_2) \end{cases} \tag{8}$$

2.2. Modeling of the Critical Depth of Cut

Based on the traditional cutting force model, indentation model, and scratch pattern, a new method considering the elastic recovery was designed and two assumptions were proposed: (1) As a rigid body, the indenter does not deform; however, it wears during the process of scratching. (2) The motion of the indenter is quasi-static. According to the characteristics of deformation, the process of scratching can be divided into three stages: the elasticity leading stage, the ductility leading stage, and the brittleness leading stage [26].

In the elasticity leading stage, the force applied on the indenter consists of an elastic restoring force, an adhesive force, and a frictional force. The elastic restoring force is a reactive force applied on the indenter caused by the elastic deformation of the part. The adhesive force between the two solids (i.e., indenter and part) is complex [27]; therefore, a simplified approach that combines the adhesive and frictional forces is used in this study. All forces are decomposed into normal and tangential forces:

$$\begin{aligned} F_{en} &= K_1 A_1 \\ F_{et} &= \mu K_1 A_1 \end{aligned} \tag{9}$$

where K_1 is the average contact pressure between the indenter and workpiece, A_1 is the projected area of the contact surface between the indenter and part ($A_1 = 3\sqrt{3}(d+d^*)^2 \tan^2\theta$), and μ is the frictional and adhesive coefficient.

The study by Son et al. [28] showed that the minimum cutting depth producing chips can be expressed as

$$d_m = R\left(1 - \cos\left(\frac{\pi}{4} - \frac{\beta}{2}\right)\right) \tag{10}$$

where β is the friction angle.

In the ductility leading stage, the force applied on the indenter consists of an elastic restoring force, frictional and adhesive forces, and cutting deformation force. The force analysis is illustrated in Figure 2. The frictional and adhesive forces consist of two parts: one caused by chip formation and one caused by elastic recovery. The cutting deformation force, which is a reaction force applied on the indenter caused by the deformation of the part, can be separated into a chip formation force and a plowing force. However, the plowing force can be ignored in this model since it is much weaker than the chip formation force [29]. All forces are decomposed into normal and tangential forces:

$$\begin{aligned} F_{dn} &= F_{dn1} - F_{dn2} + F_{dn3} \\ F_{dt} &= F_{dt1} + F_{dt2} + F_{dt3} \end{aligned} \tag{11}$$

where F_{dn1} is the normal force component caused by the elastic restoring force, F_{dn2} is the normal force component caused by the frictional and adhesive forces, F_{dn3} is the normal force component caused by the chip formation force, F_{dt1} is the tangential force component caused by the elastic restoring force, F_{dt2} is the tangential force component caused by the frictional and adhesive force, and F_{dt3} is the tangential force component caused by the chip formation force.

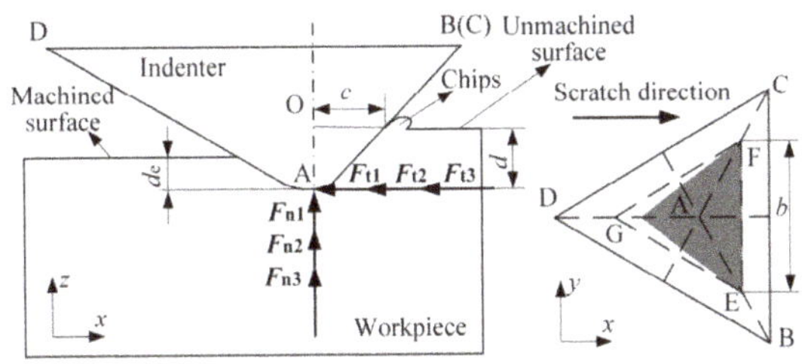

Figure 2. Force analysis model in the ductility leading stage.

The normal force components can be calculated using Equation (12).

$$\begin{aligned} F_{dn1} &= K_1 S_1 \\ F_{dn2} &= \mu(K_1 + K_2) S_2 \sin\theta \\ F_{dn3} &= K_2 S_2 \sin\theta \end{aligned} \tag{12}$$

where K_2 is the cutting deformation contact stress, S_1 is the projected area given by the shaded area in Figure 2, and S_2 is the contact area between the region of ductile deformation, which is the area from the unmachined surface to the machined surface.

The areas S_1 and S_2, respectively, are

$$\begin{aligned} S_1 &= \sqrt{3}[(d_e + d^*)\tan\alpha + (d + d^*)\tan\theta](d + d^*)\tan\theta \\ S_2 &= \frac{(d-d_e)}{2\cos\theta}[\sqrt{3}(d + d^*)\tan\theta + \sqrt{3}(d_e + d^*)\tan\theta] \end{aligned} \tag{13}$$

where d_e is the part elastic recovery depth and is equal to the height difference between the scratching and residual depths. Note that the elastic recovery depth is not constant and increases linearly as the scratching depth increases [20].

The tangential force components are

$$\begin{aligned} F_{dt1} &= K_1 S_2 \cos\theta \\ F_{dt2} &= \mu(K_1 + K_2)S_2\cos\theta + \mu K_1 S_1 \\ F_{dt3} &= \frac{F_{dn3}}{\tan\theta} \end{aligned} \tag{14}$$

The normal and tangential forces, respectively, are

$$\begin{aligned} F_{dn} &= K_1 S_1 - \mu(K_1 + K_2)S_2\sin\theta + K_2 S_2 \sin\theta \\ F_{dt} &= K_1 S_2\cos\theta + \mu(K_1 + K_2)S_2\cos\theta + \mu K_1 S_1 + K_2 S_2\cos\theta \end{aligned} \tag{15}$$

In the brittleness leading stage, the average contact pressure and the cutting deformation contact stress show a zigzag change due to crack propagation and pop-in debris.

It is difficult to control the depth of processing, but controlling the cutting force, especially the normal force, is relatively easy, no matter whether ultra-precision grinding or single point diamond cutting is used. Therefore, the cutting force model of this section can be used to control the cutting depth through the cutting force.

The dislocation will appear when the part undergoes extrusion deformation [30]. The appearance of a cleavage crack will occur when one side's tensile stress reaches the limit under the action of the applied force. The theoretical cleavage strength can be expressed by [31]

$$\sigma_c = \frac{1}{2}\sqrt{\frac{E\gamma}{a}} \tag{16}$$

where E is elastic modulus, γ is surface energy per unit area, and a is the interplanar spacing. In the scratching process, the maximum stress in the part's machined surface is located at the tip of the indenter. If the maximum stress is less than the cleavage strength, no cracks will occur on the surface or subsurface of the part. The maximum stress is [32]

$$P_0 = \frac{3}{2}K_1 \tag{17}$$

The parameter K_1 is obtained from Equation (15).

$$K_1 = \frac{F_{dn}(\mu S_2\cos\theta + S_2\cos\theta) - F_{dt}(S_2\sin\theta - \mu S_2\sin\theta)}{(S_1 - \mu S_2\sin\theta)(\mu S_2\cos\theta + S_2\cos\theta) - (S_2\cos\theta + \mu S_2\cos\theta + \mu S_1)(S_2\sin\theta - \mu S_2\sin\theta)} \tag{18}$$

The critical condition of brittle materials during the scratching process is

$$\sigma_c = \frac{3}{2}K_{1c} \tag{19}$$

where K_{1c} is the critical average contact pressure.

3. Experimental Setup

A commercial wafer of 4H-SiC single crystal, grown using the physical vapor transport method, was supplied by Shanghai Institute of Optics and Mechanics Chinese Academy of Sciences. The 4H-SiC wafer was cut to the size of 10 mm × 10 mm by a laser cutting machine and its thickness was 0.5 mm. It was measured using a scanning electron microscope (SEM), and the surface roughness was found to be less than 1 nm after a polishing treatment. All experiments conducted in this paper were conducted on the (0001) plane. A diamond Berkovich indenter with an angle between the edge plane and the centerline of 65.27° was used in this study, and a standard fused quartz sample was employed to indirectly measure the nose radius of the indenter.

A laser cutting machine (HGLaser LCC0130-CO2, HGTECH, Wuhan, China) was used to prepare the sample used in this study. A nanomechanical test instrument (TI 950 Triboindenter, Hysitron, Minneapolis, MN, USA), with load sensitivity less than 30 nN and displacement sensitivity less than 0.2 nm, was used to scratch the sample surface and record the information of the scratching depth, tangential force, normal force, and time and to obtain in situ scanning probe microscopy (SPM) images. The instrument is shown in Figure 3. Scanning electron micrographs of the scratch were generated using a foucused ion beam-scanning electron microscope (FIB-SEM) system (Helios NanoLab 600i, FEI, Hillsboro, OR, USA).

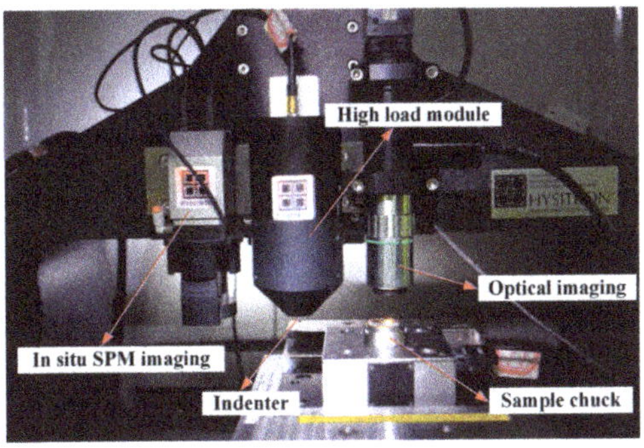

Figure 3. Core components of the TI 950 Triboindenter.

The experiments consisted of two parts, namely, the indentation and scratch experiments. In order to determine the nose radius of the indenter, nine indents were made at maximum load capacity from 20 to 180 mN at a constant interval on the standard fused quartz sample surface, and a constant loading rate was used in the process of the experiment. All experiments were performed with a 10 s holding time at room temperature.

The scratch process on the commercial wafer of the 4H-SiC single crystal included three stages: (1) the pre-scan stage, (2) the scratching stage, and (3) the post-scan stage. In the first stage, the sample surface morphology information, such as surface roughness and sample inclination angle, were obtained when the indenter was scanned on the sample surface with a constant load of 0.1 mN. In the scratching stage, the indenter load was increased linearly from 0 to 80 mN while the sample table moved at a constant rate, and the length of the scratch on the sample surface was 250 μm. In the final stage, the indenter scanned backwards with a constant load of 0.1 mN to obtain the scratched surface morphology information. The scratch test parameters are shown in Table 1. The scratched surface topography imaging was delivered by dual piezo scanners in the in situ SPM imaging system.

A FIB-SEM system was used to measure and evaluate the deformation characteristics of scratches and cracks on the samples immediately after the nanoscratching tests.

Table 1. Scratch test parameters.

Test Parameters	Unit	Values
Pre-scan/post-scan load	mN	0.1
Loading range	mN	0.1–80
Scratch length	µm	250
Scratch velocity	µm/s	4

4. Results, Analysis, and Discussion

4.1. Determination of the Indenter Nose Radius

An indirect method to compare the theoretical projected area, which is a function of R and d, with the area function acquired through nanoindentation on the standard fused quartz sample was used to determine the numerical value of R. The hardness, H, can be expressed as [33]

$$H = \frac{F_{\text{max}}}{A_{\text{p}}} \tag{20}$$

where F_{max} is the maximum load. For standard fused quartz, the hardness is 9.5 GPa. For the Berkovich indenter used in this study, $\alpha = 77.3°$, $\beta = 57.64°$, $\theta = 65.27°$, and $\gamma = 60°$. Table 2 shows the indenter height and projected area for a variety of maximum loads. Using least squares, the projected area was related to the indenter height by $A_p = 25.58 \times (123.8 + d)^2$, as shown in Figure 4. The indenter nose radius was calculated as R = 4952 nm via Equation (8).

Table 2. Indenter height and contact area for different loads.

Load (mN)	Indenter Height (nm)	Contact Area (nm^2)
20	164.5	2.2053×10^6
40	274.2	4.2905×10^6
60	379.6	6.4158×10^6
80	460.1	8.3211×10^6
100	522.9	1.0426×10^7
120	578.6	1.3032×10^7
140	625.4	1.4037×10^7
160	676.5	1.5642×10^7
180	750.4	1.8247×10^7

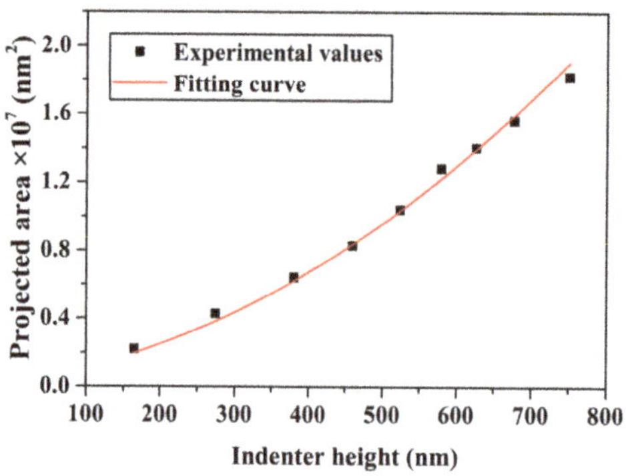

Figure 4. Relationship between the projected area and indenter height.

4.2. Analytic Surface Morphology

The surface morphology of the scratch is shown in Figure 5. It was observed in the enlarged image that material is removed but no cracks are formed in the surface at Position 1. There are cracks at the bottom of the scratch at Positions 2 to 4; these are perpendicular to the scratch motion and are the result of median crack closure and lateral crack growth due to the residual stress caused by the indenter. The size of the cracks was amplified with increasing scratching depth. Subsurface cracks were revealed with the scanning electron microscope. Therefore, the results show that the ductile–brittle transition of 4H-SiC is located before Position 2, as shown in Figure 5, and the corresponding scratch length ranges from 0 to 80 μm, with the scratching depth ranging from 0 to 120 nm.

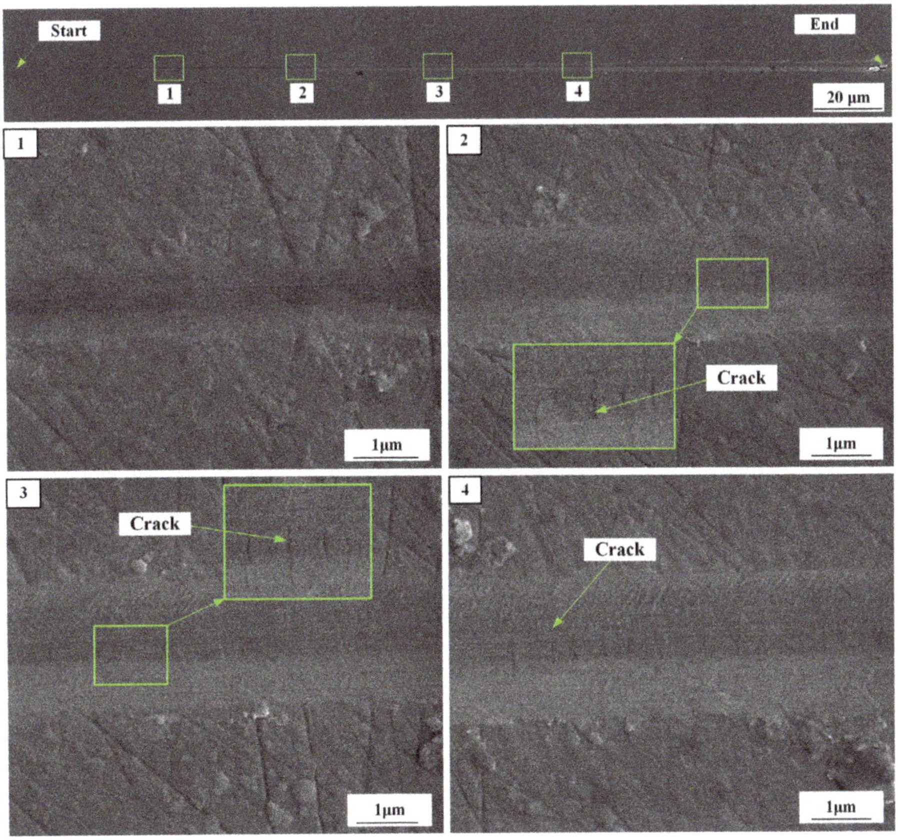

Figure 5. Full view and enlarged image of a scratch using SEM.

4.3. Comparison of the Critical Depth of Cut between Simulation and Experiments

The experimental data were obtained using a two-dimensional three-plate capacitive sensor and a TI 950 piezoelectric ceramic. The experimental results are shown in Figure 6; Figure 7. Figure 6 shows the tangential force as a function of the lateral displacement. Figure 7 shows the scratching depth as a function of the lateral displacement.

According to the theory mentioned in Section 2.2, the whole scratch can be divided into three stages: I, standing for the elasticity leading stage; II, standing for the ductile leading stage; and III, standing for the brittleness leading stage, as shown in Figure 7. The minimum scratching depth is 40 nm and the elastic recovery depth/scratching depth ratio is 0.77 through an analysis of the scratching depth versus lateral displacement curve. In the elasticity leading stage, i.e., where the scratching depth is less than 40 nm, the experimental data were plugged into Equation (9), and we received a frictional and adhesive coefficient of $\mu = 0.31$; this is much larger than the frictional coefficient, which is equal to 0.05 [34].

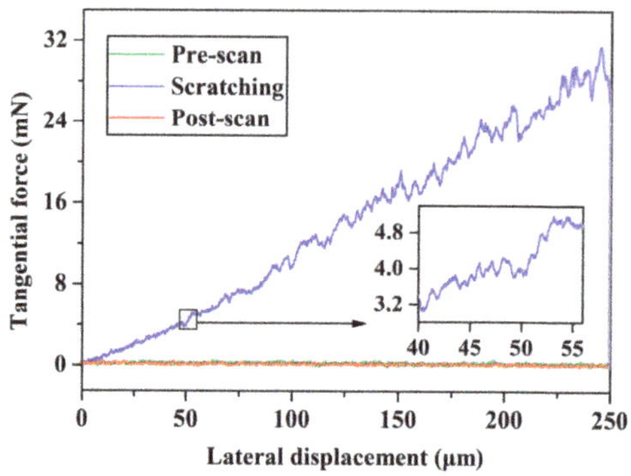

Figure 6. Tangential force as a function of lateral displacement.

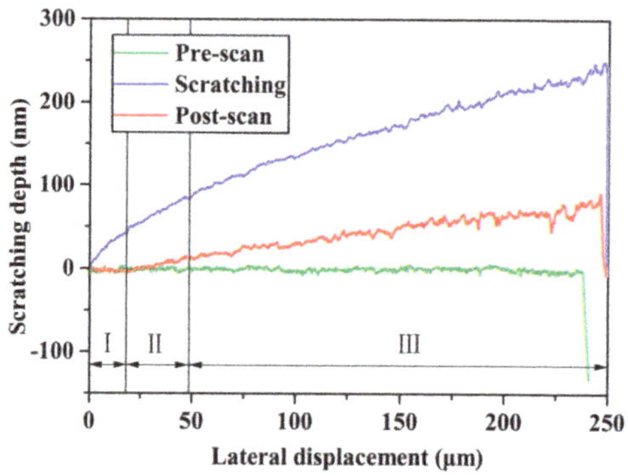

Figure 7. Scratching depth as a function of lateral displacement.

In the ductile leading stage, i.e., where $d \geq 40$ nm, the average contact pressure was computed via Equations (13) and (18) and is shown in Figure 8. The cleavage strength of silicon carbide is 26.7 Gpa [35]. The critical average contact pressure is 17.8 Gpa via Equation (19). According to the relationship between the average contact pressure and the scratching depth, the critical depth of cut of 4H-SiC was determined to be 92 nm. Extension of the crack can cause a drastic change in tangential force and the appearance of a pop-in phenomenon. The first pop-in point of the relation curve between tangential force and lateral displacement appears where the scratching depth is about 90 nm, and it is very close to the theoretical calculation results. The in situ SPM images where the lateral displacement ranged from 50 to 60 μm, including the critical depth of cut, indicate that the residual depth, when located in the critical depth of cut, is 20.8 nm, as shown in Figure 9. The two aforementioned scratch tests were repeated in order to exclude the contingency of a single-pass test. The same process was used to handle the test data, and the results are shown in Table 3. This result shows good agreement with other references, as shown in Table 4.

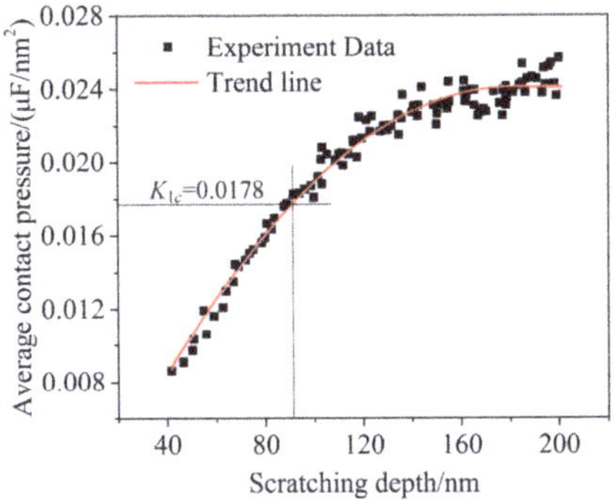

Figure 8. Average contact pressure as a function of scratching depth.

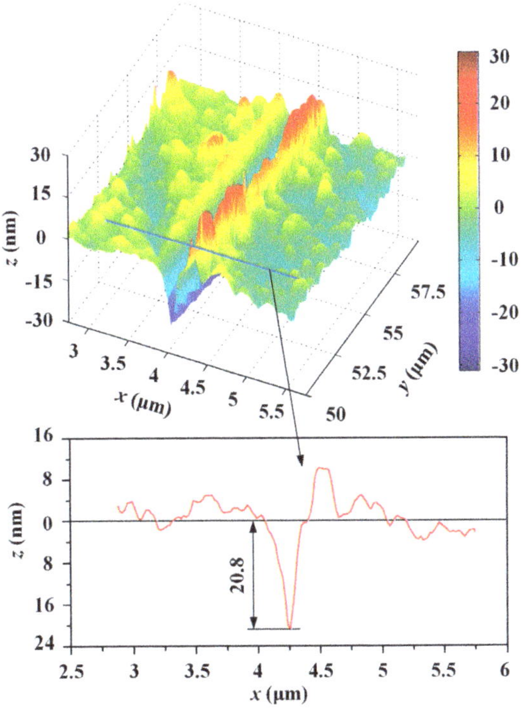

Figure 9. In situ SPM images.

Table 3. Repeated test results.

Test Number	Critical Depth of Cut (nm)
1	92
2	93
3	90
Average value	91.7

The sources of error in this study are as follows: (1) The frictional and adhesive coefficient between the indenter and sample surface is not a constant in the process of scratching when loaded linearly [41], but it was simplified to a constant in this study. (2) The wear of the indenter was ignored. (3) The impact of defects in the crystal, such as microtubules, dislocations, and stacking faults, was also ignored. (4) Although the roughness of the sample surface was less than 1 nm, it still has a considerable influence in nanoscale experiments.

Table 4. Indenter height and contact area for different loads.

Critical Depth of Cut (nm)	Material	Speed (mm/s)	Tip Radius (μm)	Refs.
75	6H-SiC	0.01	0.94	[36]
95	4H-SiC	0.001	5	[37]
<100	6H-SiC	150	0.05	[38]
70	6H-SiC	82.5	0.05	[39]
<60	6H-SiC	4.5	800	[40]

5. Summary and Conclusions

A theoretical model of the critical depth of cut of nanoscratching on a 4H-SiC single crystal with a Berkovich indenter was proposed, and a scratch test in a nanomechanical test system was conducted. The following conclusions can be drawn from this study:

(1) Based on an analysis of the nanoindentation and typical scratch model, a new model of the critical depth of cut of nanoscratching on a 4H-SiC single crystal with a Berkovich indenter was established.

(2) The radius of the Berkovich indenter nose was indirectly confirmed by a nanoindentation experiment, and the range of cracks on the scratched surface was verified by SEM images.

(3) The change in the sample surface in the scratching process was revealed through the average contact pressure. The theoretical result of the critical depth of cut at the ductile–brittle transition for a 4H-SiC single crystal was obtained; it is close to the first obvious pop-in point of the relation curve between tangential force and lateral displacement, and this result shows good agreement with other references.

Author Contributions: Conceptualization, P.C. and S.L.; Data curation, P.C. and S.L.; Funding acquisition, S.L. and Y.L.; Investigation, P.C. and S.L.; Methodology, S.L.; Project administration, Y.L.; Writing—original draft, P.C.; Writing—review and editing, S.L. and Y.L.

Funding: This work is supported by the National Natural Science Foundation of China (Nos. 51575442) and Shaanxi Provincial Natural Science Foundation (2016JZ011).

Acknowledgments: The authors are grateful to the National Natural Science Foundation of China and Shaanxi Provincial Natural Science Foundation which enabled the research on the rails.

Conflicts of Interest: The authors declare no conflict of interest.

References

1. Goel, S. The current understanding on the diamond machining of silicon carbide. *J. Phys. D Appl. Phys.* **2014**, *47*, 243001. [CrossRef]
2. Choyke, W.J.; Pensl, G. Physical properties of SiC. *MRS Bull.* **1997**, *22*, 25–29. [CrossRef]
3. Miyazaki, H.; Zhou, Y.; Iwakiri, S.; Hirotsuru, H.; Hirao, K.; Fukuda, S.; Izu, N.; Hyuga, H. Improved resistance to thermal fatigue of active metal brazing substrates for silicon carbide power modules using tough silicon nitrides with high thermal conductivity. *Ceram. Int.* **2018**, *44*, 8870–8876. [CrossRef]
4. Han, D.; Mei, H.; Xiao, S.; Dassios, K.G.; Cheng, L. A review on the processing technologies of carbon nanotube/silicon carbide composites. *J. Eur. Ceram. Soc.* **2018**, *38*, 3695–3708. [CrossRef]
5. Robichaud, J.; Guregian, J.J.; Schwalm, M. SiC optics for Earth observing applications. In Proceedings of the SPIE—The International Society for Optical Engineering, San Diego, CA, USA, 10 November 2003; Volume 5151, pp. 53–62. [CrossRef]
6. Lee, H.S.; Kim, D.I.; An, J.H.; Lee, H.J.; Kim, K.H.; Jeong, H. Hybrid polishing mechanism of single crystal SiC using mixed abrasive slurry (MAS). *CIRP Ann.* **2010**, *59*, 333–336. [CrossRef]

7. Yin, L.; Vancoille, E.Y.J.; Lee, L.C.; Huang, H.; Ramesh, K.; Liu, X.D. High-quality grinding of polycrystalline silicon carbide spherical surfaces. *Wear* **2004**, *256*, 197–207. [CrossRef]
8. Gassilloud, R.; Ballif, C.; Gasser, P.; Buerki, G.; Michler, J. Deformation mechanisms of silicon during nanoscratching. *Phys. Status Solidi* **2010**, *202*, 2858–2869. [CrossRef]
9. Nian, D.; Yiqing, Y.; Wenshan, W.; Xu, X. Analysis of grit interference mechanisms for the double scratching of monocrystalline silicon carbide by coupling the FEM and SPH. *Int. J. Mach. Tool. Manuf.* **2017**, *120*, 49–60. [CrossRef]
10. Swain, M.V. Microfracture about scratches in brittle solids. *Proc. Royal Soc. Lond. A Math. Phys. Sci.* **1979**, *366*, 575–597. [CrossRef]
11. Wasmer, K.; Ballif, C.; Gassilloud, R.; Pouvreau, C.; Rabe, R.; Michler, J.; Breguet, J.-M.; Solletti, J.-M.; Karimi, A.; Schulz, D. Cleavage Fracture of Brittle Semiconductors from the Nanometre to the Centimetre Scale. *Adv. Eng. Mater.* **2005**, *7*, 309–317. [CrossRef]
12. Lawn, B.R.; Swain, M.V. Microfracture beneath point indentations in brittle solids. *J. Mater. Sci.* **1975**, *10*, 113–122. [CrossRef]
13. Bondokov, R.T.; Lashkov, T.; Sudarshan, T.S. Influence of Structural Defects on the Polishing of Silicon Carbide Single Crystal Wafers. *Jpn. J. Appl. Phys.* **2004**, *43*, 43–49. [CrossRef]
14. King, R.F.; Tabor, D. The Strength Properties and Frictional Behaviour of Brittle Solids. *Proc. Royal Soc. Lond. A Math. Phys. Sci.* **1954**, *223*, 225–238. [CrossRef]
15. Agarwal, S.; Rao, P.V. Experimental investigation of surface/subsurface damage formation and material removal mechanisms in SiC grinding. *Int. J. Mach. Tool. Manuf.* **2008**, *48*, 698–710. [CrossRef]
16. Virkar, S.R.; Patten, J.A. Numerical Simulations and Analysis of the Thermal Effects on Silicon Carbide During Ductile Mode Micro-Laser Assisted Machining. In Proceedings of the ASME 2009 International Manufacturing Science and Engineering Conference, West Lafayette, IN, USA, 4–7 October 2009; pp. 409–416. [CrossRef]
17. Goel, S.; Luo, X.; Comley, P.; Reuben, R.L.; Cox, A. Brittle–ductile transition during diamond turning of single crystal silicon carbide. *Int. J. Mach. Tool. Manuf.* **2013**, *65*, 15–21. [CrossRef]
18. Shujuan, L.; Aofei, T.; Yong, L.; Wang, J.; Cui, D.; Landers, R.G. Analytical Force Modeling of Fixed Abrasive Diamond Wire Saw Machining With Application to SiC Monocrystal Wafer Processing. *J. Manuf. Sci. Eng.* **2017**, *139*, 041003. [CrossRef]
19. Cao, J.; Wu, Y.; Lu, D.; Fujimoto, M.; Nomura, M. Material removal behavior in ultrasonic-assisted scratching of sic ceramics with a single diamond tool. *Int. J. Mach. Tool. Manuf.* **2014**, *79*, 49–61. [CrossRef]
20. Patten, J.; Gao, W.; Yasuto, K. Ductile regime nanomachining of single-crystal silicon carbide. *J. Manuf. Sci. Eng.* **2005**, *127*, 522–532. [CrossRef]
21. Mussi, A.; Rabier, J.; Thilly, L.; Demenet, J.L. Plasticity and deformation microstructure of 4h-SiC below the brittle-to-ductile transition. *Phys. Status Solidi C* **2011**, *4*, 2929–2933. [CrossRef]
22. Xie, J.; Li, Q.; Sun, J.X.; Li, Y.H. Study on ductile-mode mirror grinding of sic ceramic freeform surface using an elliptical torus-shaped diamond wheel. *J. Mater. Process Tech.* **2015**, *222*, 422–433. [CrossRef]
23. Bifano, T.G.; Dow, T.A.; Scattergood, R.O. Ductile Regime Grinding A New Technology for Machining Brittle Materials. *J. Eng. Ind.* **1991**, *113*, 184–189. [CrossRef]
24. Xiao, G.; To, S.; Zhang, G. The mechanism of ductile deformation in ductile regime machining of 6H SiC. *Comput. Mater. Sci.* **2015**, *98*, 178–188. [CrossRef]
25. Wang, P.; Ge, P.; Li, Z.; Ge, M.; Gao, Y. A scratching force model of diamond abrasive particles in wire sawing of single crystal sic. *Mater. Sci. Semicon. Proc.* **2017**, *68*, 21–29. [CrossRef]
26. Le Houérou, V.; Sangleboeuf, J.C.; Dériano, S.; Rouxel, T.; Duisit, G. Surface damage of soda–lime–silica glasses: Indentation scratch behavior. *J. Non-Cryst. Solids.* **2003**, *316*, 54–63. [CrossRef]
27. Popov, V.L. *Contact Mechanics and Friction: Physical Principles and Applications*; Springer: Heidelberg/Dordrecht, Germany; London, UK; New York, NY, USA, 2010; pp. 56–62.
28. Son, S.M.; Han, S.L.; Ahn, J.H. Effects of the friction coefficient on the minimum cutting thickness in micro cutting. *Int. J. Mach. Tool. Manuf.* **2005**, *45*, 529–535. [CrossRef]
29. Rowe, W.B. *Principles of Modern Grinding Technology*; Elsevier Inc.: Amsterdam, The Netherlands, 2009.
30. Lawn, B.R. *Fracture of Brittle Solids-Second Edition*; Cambridge University Press: Cambridge, UK, 1993; pp. 71–86.

31. Veld, A.J.G.O.H.; Veldkamp, J.D.B. Mechanical and microscopical investigation of SiC whiskers. *Fibre Sci. Technol.* **1970**, *2*, 269–281. [CrossRef]
32. Johnson, K.L. *Contact Mechanics*; Cambridge University Press: Cambridge, UK, 1987; pp. 96–122.
33. Oliver, W.C.; Pharr, G.M. An improved technique for determining hardness and elastic modulus using load and displacement sensing indentation experiments. *J. Mater. Res.* **1992**, *7*, 1564–1583. [CrossRef]
34. Zhang, F.; Meng, B.; Geng, Y.; Zhang, Y.; Li, Z. Friction behavior in nanoscratching of reaction bonded silicon carbide ceramic with Berkovich and sphere indenters. *Tribol. Int.* **2016**, *97*, 21–30. [CrossRef]
35. Zhao, X.; Langford, R.M.; Shapiro, I.P.; Xiao, P. Onset Plastic Deformation and Cracking Behavior of Silicon Carbide under Contact Load at Room Temperature. *J. Am. Ceram. Soc.* **2011**, *94*, 3509–3514. [CrossRef]
36. Meng, B.; Zhang, F.; Li, Z. Deformation and removal characteristics in nanoscratching of 6H-SiC with berkovich indenter. *Mater. Sci. Semicon. Proc.* **2015**, *31*, 160–165. [CrossRef]
37. Ravindra, D.; Virkar, S.; Patten, J. Ductile Mode Micro Laser Assisted Machining of Silicon Carbide. In *Properties and Applications of Silicon Carbide*; Gerhardt, R., Ed.; In Tech: London, UK, 2011.
38. Bhattacharya, B.; Patten, J.A.; Jacob, J. Ductile Regime Nano-Machining of Polycrystalline Silicon Carbide(6H). In Proceedings of the 2005 Annual Meeting ASPE, Norfolk, VA, USA, 9–14 October 2005; Volume 37, pp. 403–406.
39. Jacob, J.; Patten, J.A.; Bhattacharya, B.; Couey, J.A.; Marsh, E.R. Determination of the ductile to brittle transition and critical depth of cut in 6H-silicon carbide through fly cutting. In Proceedings of the 2005 Annual Meeting ASPE, Norfolk, VA, USA, 9–14 October 2005; Volume 37, pp. 472–475.
40. Tanaka, H.; Shimada, S. Damage-free machining of monocrystalline silicon carbide. *CIRP Ann.* **2013**, *62*, 55–58. [CrossRef]
41. Xu, N.; Han, W.; Wang, Y.; Li, J.; Shan, Z. Nanoscratching of copper surface by CeO2. *Acta Mater.* **2017**, *124*, 343–350. [CrossRef]

micromachines

Article

A Nanomechanical Analysis of Deformation Characteristics of 6H-SiC Using an Indenter and Abrasives in Different Fixed Methods

Jisheng Pan [1,2,*], Qiusheng Yan [1,2], Weihua Li [3] and Xiaowei Zhang [1]

1 School of Electromechanical Engineering, Guangdong University of Technology, Guangzhou 510006, China; qsyan@gdut.edu.cn (Q.Y.); zxwei@gdut.edu.cn (X.Z.)
2 Guangdong Nanogrind Technology Ltd., Foshan 528225, China
3 Mechanical, Materials, Mechatronic and Biomedical Engineering, University of Wollongong, Wollongong, NSW 2522, Australia; weihuali@uow.edu.au
* Correspondence: panjisheng@gdut.edu.cn; Tel.: +86-20-39322418

Received: 2 April 2019; Accepted: 16 May 2019; Published: 18 May 2019

Abstract: The super-precise theory for machining single crystal SiC substrates with abrasives needs to be improved for its chemical stability, extremely hard and brittle. A Berkovich indenter was used to carry out a systematic static stiffness indentation experiments on single crystal 6H-SiC substrates, and then these substrates were machined by utilizing fixed, free, and semi-fixed abrasives, and the nanomechanical characteristics and material removal mechanisms using abrasives in different fixed methods were analyzed theoretically. The results indicated that the hardness of C faces and Si faces of single crystal 6H-SiC under 500 mN load were 38.596 Gpa and 36.246 Gpa respectively, and their elastic moduli were 563.019 Gpa and 524.839 Gpa, respectively. Moreover, the theoretical critical loads for the plastic transition and brittle fracture of C face of single crystal 6H-SiC were 1.941 mN and 366.8 mN, while those of Si face were 1.77 mN and 488.67 mN, respectively. The 6H-SiC materials were removed by pure brittle rolling under three-body friction with free abrasives, and the process parameters determined the material removal modes of 6H-SiC substrates by grinding with fixed abrasives, nevertheless, the materials were removed under full elastic-plastic deformation in cluster magnetorheological finishing with semi-fixed abrasives.

Keywords: 6H-SiC; indentation; deformation; material removal mechanisms; critical load

1. Introduction

Single crystal SiC is a third-generation semiconductor which performs well in terms of high breakdown electric field, high thermal conductivity, large band gap, high saturated electron drift velocity, low dielectric constant and good resistance to radiation damage, all of which open up completely new classes of commercial and military applications which are currently impossible or unaffordable with silicon or gallium arsenide [1,2]. The most widely used examples include their use as sources of white light for optical storage, as displays, and in radar and communication technology, and in automotive and oil exploration industries and in high-radiation environments [3,4]. Of the many different silicon carbide polytypes, 3C-SiC, 4H-SiC and 6H-SiC are the most commonly applied. Depending on the particular application, the requirements for the resulting surface of SiC substrates are stringent and often stipulate that the roughness (Ra) must be less than 0.3 nm, the wafers must have flat, smooth, and damage free surfaces; this means the machined quality directly determines the applied value and performance of the device [5,6]. However, the high hardness, stiffness, and strength of single crystal SiC make it very difficult to obtain a high-quality surface finish by mechanical methods, even with a diamond cutting tool [7].

In recent years some researchers have become interested in analyzing the mechanical properties of single crystal SiC; Yin et al. used Vickers indention and nanoindentation to carry out indentation experiments on the 6H-SiC substrates, and used grinding and polishing tests to investigate microfracture, residual damage, and surface roughness associated with material removal and surface generation [8]. Meng et al. performed a nanoscratching test on 6H–SiC with a Berkovich diamond indenter and found that the deformation and removal of 6H-SiC while nanoscratching with a sharp indenter was completely different from the single-point diamond turning (SPDT) method [9]. Goel et al. carried out a diamond turning test on single crystal 6H-SiC at a cutting speed of 1 m/s on a precision diamond turning machine to elucidate the microscopic origin of ductile-regime machining, and obtained a surface finish of Ra = 9.2 nm [10]. Yan et al. used a Berkovich nanoindenter to investigate the subsurface damage of SiC in nanoindentation tests, and found that the depth of subsurface damage was much larger than that in indentation tests, and the damaging mechanism of SiC was completely different from single crystalline silicon [11]. Xiao et al., carried out molecular dynamics (MD) simulations to investigate the atomic scale details of ductile deformation while machining of 6H SiC, and found that a taper cutting experiment on a single crystal 6H SiC wafer produced a ductile-cut surface. Moreover, a micro Raman spectroscopy of the machined surface revealed no peaks for amorphous SiC, which agreed with the MD result [12]. Lee et al. proposed a hybrid polishing technique using a mixed abrasive slurry (MAS) with colloidal silica and nanodiamonds to investigate the hybrid removal mechanism of MAS on SiC [13]. Goel et al. carried out diamond turning of single crystal 6H-SiC at a cutting speed of 1 m/s on a precision diamond turning machine to elucidate the microscopic origin of ductile-regime machining; a surface finish of Ra = 9.2 nm was obtained and a brittle–ductile transition was observed [14]. Li et al. carried out the nanoindentation test for 6H-SiC with a Berkovich indenter and set up the three-dimensional finite element simulation, the plastic deformation and cracks morphology and mechanical properties were analyzed with the maximum load P (max) [15]. Nawaz et al. investigated the nanoscale elastic-plastic deformation behavior of single crystal 6H-SiC systematically by using nanoindentation with a Berkovich indenter and observed the effect of loading rates on the critical pop-in load, pop-in displacement and maximum shear stress [16]. Pang et al. presented an experimental and numerical analysis of the deformation behavior of single-crystal 6H-SiC in nanoindentation, the results showed that classical crystal plasticity theory can be reliably applied in predicting plastic deformation of ceramic at small scales [17]. Lu et al. described the mechanical planarization machining of SiC substrates involving the Si face and C face of N-type 4H-SiC, N-type 6H-SiC, and V-type 6H-SiC with a sol-gel polishing pad, the removal mechanism of SiC substrates was investigated by nanoindentation and nanoscratching [18].

While these studies did not present concrete nanomechanical data and generally used the mechanical data in the literature to analyze material removal mechanism, nor did they analyze the material removal mechanism of single crystal SiC from a mechanical perspective whilst using abrasives in different fixed methods. In this paper, a nanomechanical test system was used to test the nanoindentation of single crystal 6H-SiC materials to obtain the hardness, modulus and loads for elastic-plastic and plastic-brittle transitions of this material. The single crystal 6H-SiC materials produced by the same factory were machined by rotational grinding with fixed abrasives, lapping with free abrasives and magnetorheological (MR) finishing of semi-fixed abrasives, respectively. Then, the mechanical properties of the materials in brittle-plastic transition under the three fixed methods of abrasives with optimal experimental conditions were calculated. On this basis, this study revealed the mechanical behavior and material removal mechanisms of single crystal 6H-SiC under the effects of abrasives by combining the morphologies of the machined surfaces and the results of nanoindentation experiments.

2. Experimental

2.1. Materials

The 6H n-type SiC Dummy Grade wafers (the C face and Si face can be identified by the primary orient flat and secondary orient flat) purchased from tankeBlue Semiconductor Co. Ltd., Beijing, China were used in this experiment. Indentation and nanoscratching tests were performed on the polished wafers and grinding and lapping experiments were carried out on the as-cut wafers and double side lapped wafers. The wafers are 300 ± 10 μm thick, the surface roughness Ra of these as-cut wafers, double side lapped wafers, and polished wafers are about 0.21 μm, 73 nm, and 0.3 nm, respectively.

2.2. Nanoindentation Experiment

Nanoindentation was carried out by utilizing the Agilent G200 Nanoindenter system (Santa Clara, CA, USA); it can also be used for nanoindentation and nanoscratch, and works as a nanomechanical microscope. This experimental process was carried out automatically, according to a set of procedures to improve the reliability and competition of the experimental data. Within a total displacement range of 1.5 mm, the indenter can generate indentations deeper than 500 μm with a displacement resolution of less than 0.01 nm and a load resolution of 50 nN, respectively, under maximum load (standard) higher than 500 mN.

In nanoindentation experiments, single crystal SiC substrates were fixed onto the carrier plate using hot melt adhesives. The authors conducted a series of static stiffness experiments on Si and C faces of 6H-SiC under loads increasing from 1 mN to 500 mN using a triangular pyramid Berkovich indenter produced by the Agilent Company (Santa Clara, CA, USA). The edge plane and edge of the indenter showed 65.3° and 77.05° angles with the center line, and the equivalent cone angle was 70.32°. In this experiment, nine points were tested under each load at 3 a.m. to avoid any influences from the surrounding environment; after which the hardness H and the elastic module E of the points with maximum depth under static stiffness were obtained directly from the software and their average values were calculated as experimental results after excluding particular data.

2.3. Fixed Abrasive Machining

The grinding experiments using fixed grinding abrasives were carried out on the DMG-6011V vertical, uniaxial and high-precision end face grinder manufactured by the Lapmaster SFT Corp. (Tokyo, Japan). On this machine tool, cup-shaped grinding wheels were used to cut and grind the workpieces axially. The contact length, contact area, and cutting angle of the grinding wheels and workpieces were constant and half of the machined workpieces were always outside the grinding wheels. Furthermore, by using online thickness measuring devices, crystal plates were ground precisely and were not affected by any wear in the grinding wheels (Figure 1). The movement of the spindle of this machine tool was set at 0.01 μm per unit for its minimum rate of movement is 0.01 μm/sec, while the grinding thickness was controlled at 0.1 μm. The experiments utilized #325 metal-bonded and #8000 ceramic-bonded diamond grinding wheels, with deionized water as a coolant. The experimental parameters are shown in Table 1.

Table 1. Grinding conditions used in experiments.

Process No.	Grinding Wheel Type (Abrasive Grain Size)	Feed Rate f_s (μm/s)	Workpiece Speed n_w (rpm)	Wheel Speed n_s(rpm)
No.1	325# (45μm)	5	151	1800
No.2	325# (45μm)	0.1	151	1800
No.3	325# (45μm)	0.1	151	3200
No.4	8000# (1.6μm)	0.1	151	3200

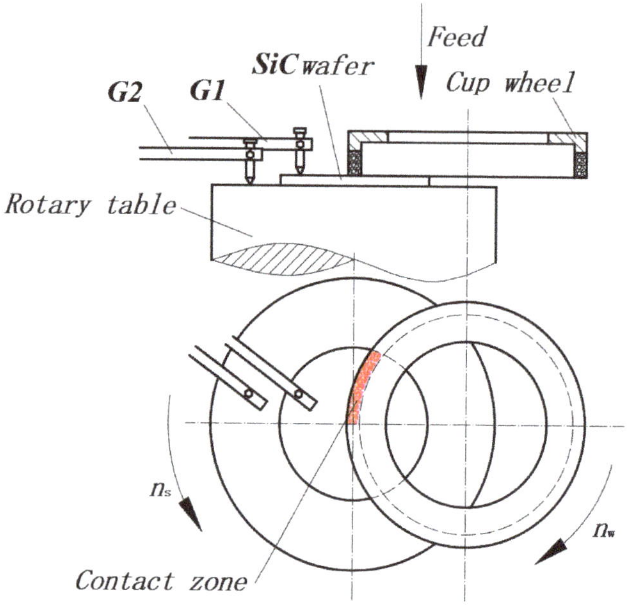

Figure 1. Principle of workpiece rotation grinding.

2.4. Free Abrasive Machining

The lapping experiments using free abrasives were carried out on the KD15BX (Dongguan KIZI Precision Lapping Mechanical Manufacture Co., LTD., Dongguan, China) precision plane grinder. This machine includes precision lapping plates, a trim ring, ceramic plates, cushions, clump weights, a retaining hook structure, a precision dressing machine, and a supply system of slurry. After being dressed for flatness, the lapping plates had a total thickness variation (TTV) of less than 10 μm and the parallelism of the upper and lower sides of the ceramic plates was less than 2 μm. The workpiece was fixed onto the ceramic plates with paraffin and faced to the lapping plate, while the trim ring was fixed outside the round ceramic plates. Clump weights were placed over and isolated with the ceramic plates by cushions. The pulley of the retaining hook structure was tangential with the outer cylinder of the trim ring so that the trim ring and ceramic plates could rotate due to friction between the retaining hook structure and the lapping plates. The experiments were carried out on the cast iron lapping plate and ceramic lapping plates using W14 and W1.5 diamond abrasives for 5 min as the lapping plates were rotating at 80 rpm. The mass fraction of abrasives in the slurry, flows of slurry, and lapping pressure were 4 wt.%, 15 mL/min, and 30 kPa, respectively.

2.5. Semi-fixed Abrasive Machining

The MR finishing experiments with semi-fixed abrasives were carried out using the experimental equipment for MR plane finishing that was developed by the laboratory (Figure 2). This equipment utilized the servo motion of CNC milling machines and was controlled in the four directions (X, Y, C1 and C2) needed for the polishing process via programming. To analyze the material removal mechanism of single crystal SiC substrates under MR processing, N35 cylindrical permanent magnets (20 mm × 15 mm) with a flat bottom were arranged circumferentially on the polishing plates in a 135 mm diameter in the same direction as the magnetic poles to form a cupped polishing circle. The single crystal 6H-SiC substrates were 2 inches in diameter and were faced directly in the center of the magnetic pole by adjusting movement in the X and Y directions. The distance between the lower surface of the substrates and the upper surface of the polishing plates was set at 0.8 mm by controlling the Z axis. The workpiece was static, while the polishing plates rotated at C1 = 200 r/min to machine the workpiece for 35 min, thus producing arc polishing belts on the workpiece. In the experiment, water-based MR fluids mixed with certain proportions of abrasives were used as MR polishing fluids. These MR polishing fluids consisted mainly of W3.5 carbonyl iron powders (4 wt.%), deionized water (88 wt.%), W3.5 diamond powders (4 wt.%), and stabilizer (4 wt.%).

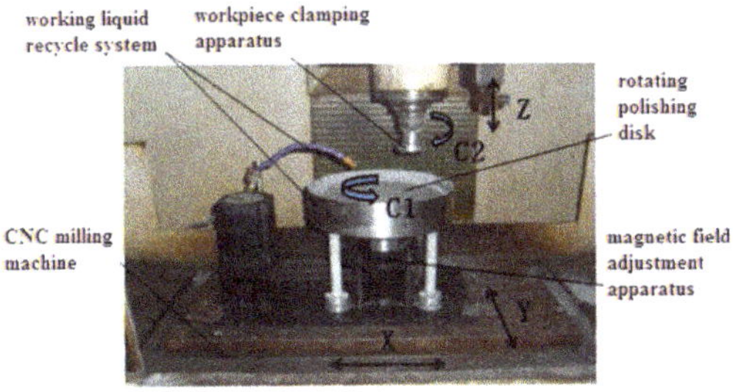

Figure 2. Experimental apparatus for cluster MR effect in plane polishing.

The surface roughness was measured with a MarWin XT20 (Mahr, Goettingen, Germany) surface roughometer, and the surface morphology was analyzed by OLS4000 (Olympus Corporation, Tokoy, Japan) laser confocal microscopy and a scanning electron microscope (SEM) (Hitachi, Ltd., Tokoy, Japan).

3. Results and Discussion

3.1. Results and Theoretical Analysis of this Nanoindentation Experiment

The experiments were carried out on the C and Si faces of polished 6H-SiC substrates by utilizing the static stiffness method under 1 mN, 2 mN, 5 mN, 10 mN, 20 mN, 50 mN, 100 mN, 200 mN, 300 mN and 500 mN loads. Each experiment was repeated nine times in a 3 × 3 array. The surface morphologies of the 6H-SiC substrates obtained from the experiment are shown in Figure 3 and the hardness and elastic moduli of the C and Si faces are shown in Table 2.

The results indicate that single crystal SiC substrates showed the obvious effects of indentation size, whilst due to the influences of surface roughness, oxide layers on the surfaces and elastic-plastic deformation of the workpieces, the hardness and elastic modulus gradually increased as the loads and depth of indentation increased in a certain range. Afterwards, due to the soft base of the hot melt adhesives used as test patches, the hardness and elastic modulus gradually decreased as the loads and depth of indentation increased. The hardness of static stiffness of C and Si faces under 500 mN loads were 38.596 Gpa and 36.246 Gpa, respectively, which were basically consistent with the experimental result (38 Gpa) of Chen et al. [19]. Moreover, the elastic moduli of the static stiffness of C and Si faces under 500 mN loads were 563.019 Gpa and 524.839 Gpa, respectively, which were slightly larger than the result (448 Gpa) of Mehregany et al. [20]; this was probably due to the soft base of the hot melt adhesives. Therefore, the C and Si faces of single crystal 6H-SiC showed slightly different mechanical properties and demonstrated obvious anisotropy. Furthermore, the hardness and elastic moduli of the C face were larger than the Si face, which indicated that the C face was more difficult to machine than the Si face.

Table 2. Hardness and elastic modulus of Single crystal 6H-SiC substrates by quasi-static indentation.

Hardness/ Elasticity Modulus \ Loads	1 mN	2 mN	5 mN	10 mN	20 mN	50 mN	100 mN	200 mN	300 mN	500 mN
Hardness of C face (Gpa)	47.867	55.418	55.825	52.596	50.514	46.745	43.741	41.435	39.861	38.596
Elasticity modulus of C face (Gpa)	538.075	581.841	627.822	618.019	612.893	596.833	576.843	567.615	562.4	563.019
Hardness of Si face (Gpa)	48.211	46.381	50.788	48.446	45.693	42.392	39.651	37.315	36.311	36.246
Elasticity modulus of Si face (Gpa)	533.033	554.146	614.533	601.814	589.13	570.998	548.221	535.773	525.946	524.839

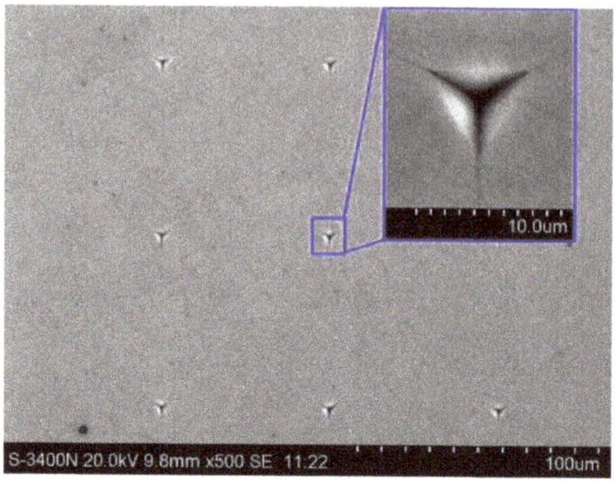

Figure 3. Surface topography of nanoindentation.

By using the static stiffness method, the load-depth curves between loads and indentation depth of the indenter are shown in Figures 4 and 5. Here, at 1.941 mN, a slight pop-in appeared on the C surface of SiC (considering the effects of the soft base of hot melt adhesives, the indentation depth was not used as a reference standard), while a micro pop-in was found on Si surface of 2.710 mN. Therefore, pop-ins under small loads were considered to be the turning point where materials changed from elastic to plastic deformation. Similarly, under higher loads, micro pop-ins appeared on the C and Si surfaces of SiC in 385.362 mN and 448.217 mN, respectively, showing that obvious brittle fractures occurred to the SiC materials under these loads.

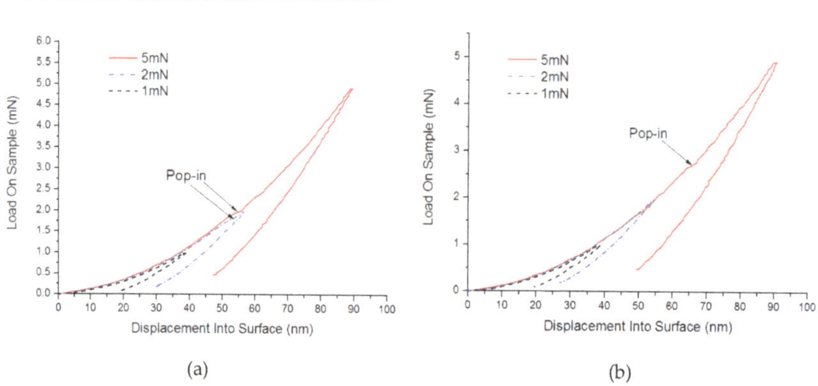

Figure 4. Test results of Single crystal 6H-SiC substrates under small loads (**a**) Load and depth curve in small loads of C face and (**b**) Load and depth curve in small loads of Si face.

In the nanoindentation experiment, a single ideal diamond grinding grain acted perpendicularly onto the single crystal SiC substrates. For brittle materials, as the load increased, they went through stages such as elastic and plastic deformation, and brittle fracture. Elastic deformation was mainly shown as an increase or decrease of the interatomic distance where elastic recovery occurred after unloading, while plastically deformed materials cannot recover to their original shapes after unloading. Under the effect of plastic deformation, plastic flows and plastic budges were generated on the materials which caused the easy formation and accumulation of defects such as fault and slippage, and because these stresses were beyond the minimum stress required by crack extension, brittle fractures caused the materials to produce microcracks which travelled inside the materials. The transition from elastic to plastic deformation and then to brittle fracture was attributed to the pop-ins of critical stress and strain on the materials. Therefore, the effective evaluation and test of the critical stress and strain of the materials provided an important basis for analyzing the material removed.

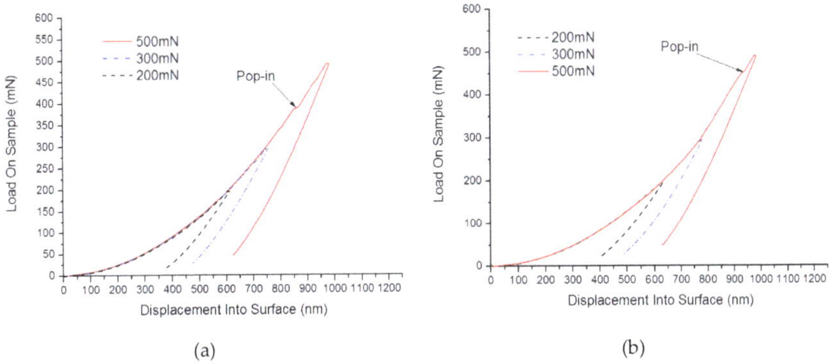

Figure 5. Test results of Single crystal 6H-SiC substrates under large loads (**a**) Load and depth curve in large loads of C face and (**b**) Load and depth curve in large loads of Si face.

According to the simple contact load model proposed by Stevanovic for predicting elastic, elastic-plastic, and plastic deformations [21]:

$$F_o = \frac{9\pi^3}{16} H^3 \left(\frac{\rho}{E}\right)^2. \quad (1)$$

When the force on surfaces of the material is $F \leq \frac{1}{15}F_o$, it is considered that pure elastic deformation occurred, and elastic-plastic deformation occurs when the force is $\frac{1}{15}F_o < F < 15F_o$; moreover, pure plastic deformation appears when the force is $F \geq 15F_o$. In the Equation, H, ρ, and E represent the hardness of the materials, the radius of curvature of grinding grains on the materials, and the elastic moduli of the materials, respectively.

In Equation (1), the hardness and elastic moduli of single crystal SiC were obtained through the nanoindentation experiment, while the radius ρ of the curvature of the Berkovich indenter was unknown. When $F = 15F_o$ was used as the critical load for the plastic transition of single crystal 6H-SiC, then

$$F_C = 15\frac{9\pi^3}{16} 38.5^3 \left(\frac{\rho_C}{560}\right)^2 = 1.941 \text{ mN} \quad (2)$$

$$F_{Si} = 15\frac{9\pi^3}{16} 36^3 \left(\frac{\rho_{Si}}{524}\right)^2 = 2.71 \text{ mN}. \quad (3)$$

By solving Equations (2) and (3), $\rho_C = 0.2$ mm and $\rho_{Si} = 0.3$ mm. As a single crystal SiC is very hard, when the Berkovich indenter was pressed into the surfaces of carbon and silicon in batches, it was worn, which further influenced the precision of the measuring results. The silicon surface was trimmed and calculated according to ρ_C value, so theoretically the critical load for the plastic transition of the silicon surface under the same conditions was

$$F_{Si'} = 15\frac{9\pi^3}{16} 36^3 \left(\frac{0.2}{524}\right)^2 = 1.77 \text{ mN}. \quad (4)$$

Therefore, when the forces on the surfaces of single crystal 6H-SiC materials were $F_c \leq 8.63$ μN and $F_{si} \leq 7.87$ μN, pure elastic deformation occurred, but if the forces were 8.63 μN $< F_c <$ 1.941 mN and 7.87 μN $< F_{si} <$ 1.77 mN, elastic-plastic deformation appeared, and if the forces were $F_C \geq 1.941$ mN and $F_{si} \geq 1.77$ mN, pure plastic deformation occurred.

When the materials were processed by plastic removal, plastic flows which did not result in cracks were observed on the surface layers, but because the materials hindered the dislocation guide, while the loads increased constantly, many dislocations accumulated at a point to form a dislocation pile-up group. Once the force reached a certain limited value, the dislocation pile-up group induced the generation and extension of cracks, which turned into brittle removal [22], and therefore the critical loads needed for the materials to produce cracks were those for the brittle transition of materials, and the critical loads were mainly determined by mechanical properties of the materials. These mechanical properties included the hardness H, the fracture toughness Kc, and elastic module E of the materials. Based on the model of indentation fracture mechanics [23,24], the critical load P* for brittle materials changing from plastic deformation to brittle fracture was expressed as

$$P^* = 54.5(\alpha/\eta^2\gamma^4)(K_c{}^4/H^3) \tag{5}$$

where α (for ordinary Victorinox indenter, α = 2/π), η and γ ($\eta \approx 0.6, \gamma \approx 0.1$) are constants. K_c and H represent fracture toughness (generally, K_c is 1.9 Mpa m1/2 for single crystal SiC [25]) and hardness of materials. The theoretical critical loads for the brittle fractures of the C and Si faces of single crystal SiC are defined below respectively.

$$P^*{}_C = 54.5((2/\pi)/(0.6^2 0.1^4))(1.9^4/38.5^3) = 366.82 \text{ mN} \tag{6}$$

$$P^*{}_{Si} = 54.5((2/\pi)/(0.6^2 0.1^4))(1.9^4/36^3) = 448.67 \text{ mN}. \tag{7}$$

Note that the results of this theoretical calculation were consistent with those obtained in the nanoindentation experiment and can be used as basic parameters for analyzing the removal of materials machined by different fixed abrasives.

3.2. The Results of Grinding with Fixed Abrasives and Materials Removal Analysis

Since the grains in the grinding experiment were always fixed onto the grinding wheels, it is known as machining with fixed grinding grains (Figure 6). Figure 7 presents the surface SEM morphologies of ground 6H-SiC substrates. Note also that the diameter, feed, and linear speed of the grinding wheels influences the surface finish, so the morphologies of the machined surfaces obviously changed and the methods for removing material were also different. Under the No.1 process condition, most single crystal SiC was removed through brittle fracture and small amounts of plastic removal. After decreasing the feed of the grinding wheels in the No.2 process, brittle removal dominated and there was more plastic removal. As process No.3 shows, the linear speed of the grinding wheels affected the material removal modes, and the faster the linear speed, the larger the proportion of plastic removal. Although same process parameters were used in the No.3 and No.4 conditions, No.4 resulted in completely different effects where materials were mainly removed via the plastic mode because the grains in the grinding wheels had different diameters and densities.

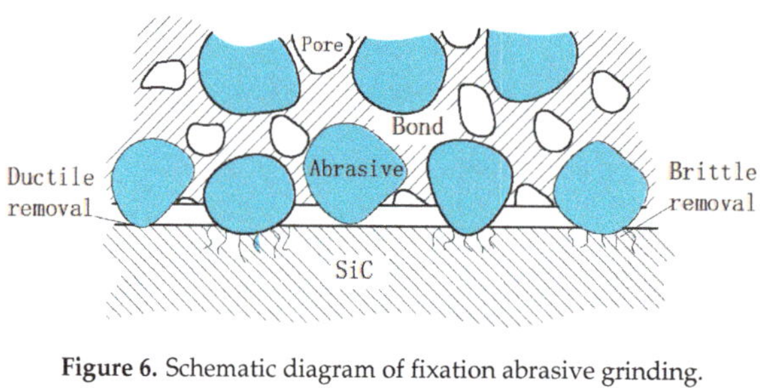

Figure 6. Schematic diagram of fixation abrasive grinding.

Figure 7. Morphology of Single crystal 6H-SiC substrates ground with different grades of grit (**a**) No.1 (Ra 0.303 μm), (**b**) No.2 (Ra 0.029 μm), (**c**) No.3 (Ra 0.015 μm) and (**d**) No.4 (Ra 0.002 μm).

Many studies have shown that the amount of material removed by grinding is related to the cutting depth d_c of the grit, which in turn is related to the properties of materials and the maximum thickness of undeformed substrate $h_{\max}$. When the maximum thickness of undeformed substrate $h_{\max}$ is less than the critical grit cutting depth d_c, the plastic domain of grinding is achieved [26].

When grinding takes place the brittleness H/K_c of materials is the principal factor influencing the brittle-plastic removal of materials.

Bifano et al. [26] put forward the critical grit cutting depth d_c for the brittle-plastic transition of material removal modes in grinding.

$$d_c = 0.15(\frac{E}{H})(\frac{K_c}{H})^2. \tag{8}$$

In Equation (9), E, H, and K_c stand for the elastic module, hardness, and fracture roughness of materials, respectively, so by substituting the experimental results above into Equation (8), the authors found that the critical grit cutting depth for C and Si faces of single crystal 6H-SiC were 5.3 nm and 6.1 nm, respectively.

During grinding, the maximum thickness of undeformed substrate $h_{\max}$ denotes the maximum depth of grinding grit cutting into the workpiece. Apart from the size of the grit and the feed rate of the grinding wheels passing over the workpiece, the maximum thickness of undeformed substrate was also related to factors such as the speed at which the workpieces and grinding wheels rotate and the size of grinding wheels. During grinding realized by the self-rotation of the workpieces, suppose that the speed at which the grinding wheel and workpieces rotate, and the feed and radius of the grinding wheels are represented by n_s, n_w, f and R, respectively. L_w and W stand for the circumference and tooth width of layers of grinding material of the cupped grinding wheels. Furthermore, C, F_v

(when the diamond density was 3.25 g/cm^3, $F_v = 0.25C$) and r_m indicate the concentration of grinding materials of diamond grinding wheels, the volume fraction of the layers of material in the grinding wheels, and the radius at any point on the surface of the workpieces, respectively. According to the study by Shang [27], the maximum thickness of undeformed substrate $h_{\max}$ was:

$$h_{\max} = 2.239R(\frac{4fr_m n_w}{L_w W C n_2^2})^{0.4}. \tag{9}$$

In this experiment, the radius R of the grinding wheels and the tooth width W of the layers of grinding material in the cupped grinding wheels were 100 mm and 3 mm, respectively. The r_m value is the position of the edges of the workpiece (for a 2-inch diameter workpiece, r_m was 25.4 mm). Moreover, the value of concentration C in Equation (9) can be obtained through simple geometric relationships, as shown in the following equation [28]:

$$C = \frac{4f}{d_g^2(\frac{4\pi}{3v_d})^{2/3}}. \tag{10}$$

In Equation (10), d_g, v_d, and f represent the equivalent spherical diameter of diamond particles, the volume fraction of diamonds in the grinding wheel, and the fraction of diamond particles that actively cut when grinding, respectively. The grinding wheel used in the present study had a density of 100, or in other words, the volume fraction v_d was 0.25. To obtain the value of C, it was assumed that only one-half of the diamond particles on the wheel surface was actively engaged in cutting [28], or the value of f was equal to 0.5.

By substituting the experimental conditions and parameters in Table 1 into Equations (9) and (10), the maximum thicknesses of undeformed substrates from processes No.1 to No.4 were 47.59 nm, 9.95 nm, 7.56 nm, and 0.43 nm, respectively. It is obvious that the maximum thicknesses of undeformed substrates from No.1 to No.3 were larger than the critical grit cutting depth dc, and therefore the materials were removed by brittle fractures. As the maximum thickness of undeformed substrates decreased the proportion of plastic removal on the surfaces gradually increased, however, the maximum thickness of the undeformed substrate in process No.4 was smaller than the critical grit cutting depth dc, so the material removal modes were completely transformed into plastic flows.

3.3. The Lapping Results of Free Abrasives and Analysis of Material Removal

Free abrasives in the form of slurry which covered the entire surface were used for lapping the workpieces. These abrasives created typical two-body and three-body friction, as shown in Figure 8. In three-body friction the abrasives are dispersed and move freely between the workpiece and the lapping plate so they simultaneously rub against the workpiece and the lapping plate; this meant that the surface materials of the workpiece are removed through surface rolling and scratching resulting in a uniform and lustrous surface consisting of countless micro broken pits. In two-body friction the abrasives become embedded into the lapping plates and only rubbed against the workpiece; this means the loading pressure is transferred directly onto the abrasives. In this process, the abrasives cut deeper into the workpiece than during three-body friction, and the workpiece materials are mainly removed through ploughing and micro-cutting modes. This indicates that the material removal rate is higher under two-body friction [29].

Figure 9 shows the surface morphologies of the cast iron and ceramic plates lapped using W14 and W1.5 diamond abrasives; the surfaces of the workpieces consist of countless micro broken pits whose size depends on whether W14 or W1.5 abrasives used. This means that material was mainly removed by a larger number of brittle fractures which caused debris to break away from the material and leave many broken pits on the surface of the workpiece; this is where three-body friction occurred. The larger the diameter of the abrasives, the less abrasives per unit area and therefore a greater force was exerted on single abrasives under the same loads, and the larger the material removed, the rougher the surface was.

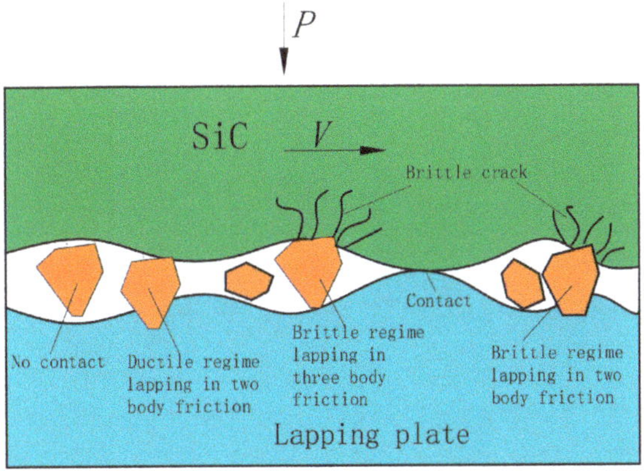

Figure 8. Three-Body and two-body abrasion.

(a) (b)

Figure 9. Morphology of single crystal 6H-SiC substrates after lapping. (**a**) After lapping by W14 diamond. (**b**) After lapping by W1.5 diamond.

In lapping, the space h_{wp} between the workpieces and the lapping plates was mainly determined by the maximum diameter $d_{\max}$ of abrasives in the space, so abrasives with diameters larger than h_{wp} did most of the lapping. The distribution laws of the diameters of abrasives met the normal probability density function [30]. Suppose that abrasives with diameters larger than D50 (D50 of W14 and W1.5 was 11 μm and 1.2 μm) did most of the lapping, then abrasives filled all of the space between the workpieces and lapping plates. Based on indentation theory, the contact deformation when abrasives were pressed into the workpieces and lapping plates is shown in Figure 10. Assume that H and F_w indicate the hardness of the workpieces and the average pressure on single abrasive, then [31],

$$h_w = \sqrt{\frac{F_w k_\beta}{H}}. \tag{11}$$

K_β is the coefficient relating to the shape of the grinding grains. Suppose that the grinding grains are ideal spheres with a radius d_w, the actual contact area between a single abrasive and the workpieces may be calculated as:

$$S_A = \pi\left(\left(\frac{d_w}{2}\right)^2 - \left(\frac{d_W}{2} - h_W\right)^2\right). \tag{12}$$

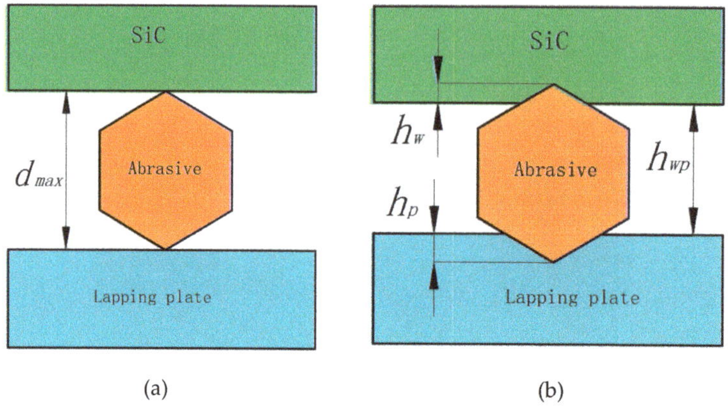

Figure 10. Contact state diagram of the abrasives, workpiece, and lapping plate. (**a**) Ideal contact state. (**b**) Actual contact state.

In accordance with the equilibrium condition of forces, the following equation is obtained.

$$pS = F_W K_b (4S/\pi d_w^2)\pi\left(\left(\frac{d_w}{2}\right)^2 - \left(\frac{d_W}{2} - h_W\right)^2\right) \quad (13)$$

where p, S, and K_b show the lapping pressure, the workpiece area, and the distribution coefficient of abrasives between the workpieces and lapping plates, respectively. When the abrasives were distributed closely, the coefficient is 1. Combining Equations (11) and (13), the following equation was obtained.

$$F_W = \sqrt{pd_W H/4K_b K_\beta}. \quad (14)$$

When K_β and K_b were 1 and the nanoindentation results and grinding parameters were substituted [31], the forces of effective W14 abrasives on C and Si faces were 1723 mN and 1728 mN, separately, and the forces of effective W1.5 abrasives on C and Si faces were 569 mN and 588 mN, respectively. Obviously, where the abrasives completely filled the gaps between workpieces and lapping plates, the forces for pressing abrasives in workpieces under two process conditions were beyond the theoretical critical load for brittle fractures of the single crystal SiC substrates. However, many abrasives between the workpieces and lapping plates could not be involved in lapping, thus causing larger actual forces for pressing abrasives into workpieces. Therefore, in two process conditions such as three-body friction and rolling of abrasives, pure brittle fractures occurred on the surface of single crystal SiC, and the rolling and fragile broken accumulation removed workpiece materials and formed uniform and smooth surfaces consisting of countless micro broken pits.

3.4. The Cluster MR Finishing Results of Semi-fixed Abrasives and Analysis of Material Removal

In cluster MR plane finishing, small magnetic bodies are embedded into polishing plates made of diamagnetic materials according to the cluster principles. When MR fluids mixed with abrasives are poured onto the polishing plates, the upper magnetic bodies of the MR fluids are lined as chain structures along the direction of magnetic lines of forces to solidify and form semi-fixed Bingham viscoelastic micro grinding heads. This array of micro grinding heads regularly formed as polishing pads. When the workpiece is close to the polishing plates and moves along it, positive pressures and shear forces are generated in the contact areas such that the surfaces materials of the workpieces are removed. Owing to the semi-fixed and soft-constraint of viscoelastic MR fluids for the abrasives, the material on the surfaces of the workpieces are removed flexibly without damaging and scratching, so the polishing process is very efficient and results in super smooth surfaces [32].

Figure 11 shows the surface SEM images at the entrance and exit of the micro grinding heads in the polishing belts on the 6H-SiC substrates obtained from the experiment. Note the deep elastic-plastic grooves where the arc polishing belts entered the surfaces of single crystal 6H-SiC; these grooves were deeper at the entry edge and then became shallower towards the interior, and since there were no grooves where the arc polishing belts exited, the surfaces were smoother and flatter. However, some pits and pores generated by the lapping process remained but even though the polishing belts left deep grooves, there were no brittle removal scratches.

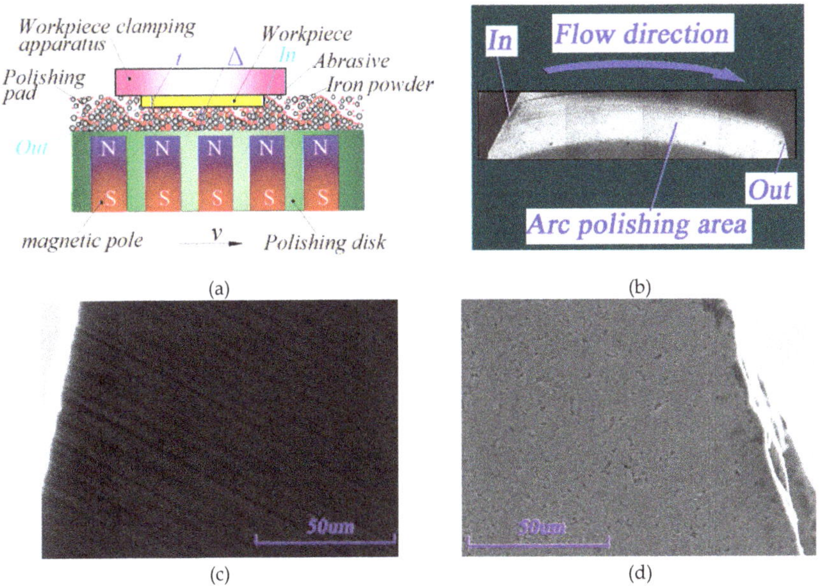

Figure 11. Scanning electron microscope (SEM) morphology of Single crystal 6H-SiC wafer polishing belt. (**a**) Schematic diagram of polishing, (**b**) schematic diagram of the polishing belt, (**c**) SEM morphology of entrance and (**d**) SEM morphology of exit.

The removal rate of material during polishing is generally related to the mechanical properties of the polishing pads. Yong's modulus of MR elastic polishing pads was ~1 Mpa(~2G/), which was much smaller than traditional polishing pads (~50–100 Mpa) [33]. Therefore, the abrasives constrained by magnetic connection were dislocated and deformed by the cutting forces, causing the accommodated-sinking effect. This was why the forces exerted by the abrasives onto the workpieces was small and the materials were removed from the surface of workpieces under elastic-plastic deformation.

Furthermore, in accordance with the Preston equation, the rate of removing MR material was proportional to the pressure on the surface of workpieces and the relative speed. As Figure 11a shows, the polishing pressure P_F of the MR micro grinding heads on the workpieces was a complex parameter which included hydrodynamic pressures, and pressures produced by the MR effects and liquid buoyancy. Moreover, the pressures produced by MR effects consisted of magnetizing and magnetostrictive pressures, and since MR fluids are non-compressible, the magnetostrictive pressures in the magnetic fields induced by changes in volume were approximately zero, the expression for polishing pressures is

$$P_F = P_d + P_g + P_m \tag{15}$$

where P_d and P_m represent the hydrodynamic pressure and the pressure produced by MR effects of the MR fluids, respectively. Furthermore, P_g stands for the buoyancy of MR fluids, but it was ignored in the calculation because it was much less than P_d and P_m.

According to the studies of Feng et al. [34], the pressure produced by the MR effects of spherical magnetic particles in MR fluids due to external magnetic fields was calculated using the following equation.

$$P_m = \frac{3\varphi\mu_0(\mu - \mu_0)}{\mu + 2\mu_0} \int_0^{H_m} H_m dH_m \tag{16}$$

where μ_0, H_m, μ and φ represent the magnetic inductivity of a vacuum, the strength of the external magnetic fields on the surface of the workpieces, the magnetic inductivity of magnetic particles, and the proportion of magnetic particles in the MR fluids, respectively. Obviously, the pressure produced by the MR effects was directly proportional to the magnetic field strength H of the external magnetic fields on the surface of the workpieces.

In cluster MR polishing, because the workpiece and the polishing plates are parallel, there is no wedge pop-in, so in theory, dynamic pressures could not be formed and there were only the pressures produced by the MR effects. However, where the micro grinding heads entered the workpieces from the bottom surface of tool heads, a pop-in of height existed, and the dynamic pressures existing at the edges of polishing belts will meet the following equation [35].

$$\frac{dP_d}{dx} = 6\eta v \frac{h - h_0}{h^3} \tag{17}$$

where η and h_0 denote the initial viscosity of the MR fluids and the distance (i.e., h_0 is the machining gap Δ) from the polishing plates to the surfaces of workpieces, respectively, then h represents the height from the polishing plates to the surfaces of tool heads. If the thickness of workpieces is t then $h = \Delta + t$.

In machining, the material removal modes are only related to the maximum force of the grinding grains, which means the forces of grinding grains in the machining gaps can be transmitted and the forces of each grinding grain and carbonyl iron powder grain are basically consistent. Moreover, the number of grinding grains and carbonyl iron powder grains in the machining gaps in the same chain is Δ/d_w so the abrasive forces at the entrance and the exit of the polishing belts in contact with the workpieces can be calculated approximately as:

$$F_{\text{in}} = \frac{\Delta}{d_w}(\frac{3\varphi\mu_0(\mu - \mu_0)}{\mu + 2\mu_0}H_m + 6\eta v \frac{t}{(\Delta + t)^3})\ \pi\left(\frac{d_w}{2}\right)^2 \tag{18}$$

$$F_{\text{out}} = \frac{\Delta}{d_w} \frac{3\varphi\mu_0(\mu - \mu_0)}{\mu + 2\mu_0} H_m\ \pi\left(\frac{d_w}{2}\right)^2. \tag{19}$$

When the data in Table 3 were substituted into Equations (18) and (19), the theoretical pressures of abrasives at the entrance and exit of workpieces were computed as 1514.27 μN and 3.47 μN, respectively; this meant they were in the mechanical range for single crystal SiC producing elastic-plastic and elastic deformations and therefore single crystal SiC materials were removed in full elastic-plastic mode, thus realizing machining without any sub-surface damage.

Table 3. Calculating Parameters.

Parameter	Value
Thickness of workpiece t (mm)	0.3
Machining gap Δ (mm)	0.8
Abrasive diameter d_w (μm)	2.8
The initial viscosity of the MR fluid η (Pa·s)	0.5
The magnetoconductivity of magnetic particles μ	2000
Vacuum permeability μ_0	1
Magnetic field intensity H_m (Gs)	2000
Speed of polishing disk v (m/s)	1.27
The proportion of magnetic particles in the MR fluid φ	0.33

4. Conclusions

The results of nanoindentation experiments under static stiffness showed that the hardness of C and Si faces of single crystal SiC under 500 mN loads was 38.596 Gpa and 36.246 Gpa, and the elastic moduli of the C and Si faces were 563.019 Gpa and 524.839 Gpa, respectively. According to the experimental results, the theoretical critical loads for the plastic transition of C and Si faces of single crystal SiC were calculated as 1.941 mN and 1.77 mN and the theoretical loads for their brittle fractures were 366.8 mN and 488.67 mN, respectively. These results basically coincided with the experimental results of nanoindentation.

In grinding based on the rotation of workpieces using fixed abrasives, the critical grit cutting depth of abrasives on the C and Si faces of single crystal 6H-SiC were 5.3 nm and 6.1 nm, separately. When the maximum thickness of undeformed substrate relating to the processing parameters was less than the critical grit cutting depth, single crystal SiC material was removed by plastic deformation, and when the maximum thickness of undeformed substrate was larger than the critical grit cutting depth, the removal mode changed to brittle removal. Moreover, as the maximum thickness of undeformed substrate increased the proportion of brittle removal on the surfaces gradually increased.

The forces of lapping materials on C and Si faces by using free abrasives (W14) were 1723 mN and 1782 mN, while those obtained using free abrasives (W1.5) were 569 mN and 588 mN, respectively. These forces were larger than the theoretical critical loads for the brittle fractures of single crystal SiC substrates. With the three-body friction motion of abrasives, single crystal SiC material was rolled, broken, and removed under pure brittle fracture.

In the cluster MR finishing experiments where semi-fixed abrasives were used, the theoretical pressure of abrasives at the entrance and exit of the polishing belts were 1514.27 μN and 3.47 μN, respectively. These pressures were in the range for elastic-plastic and elastic deformation of single crystal SiC. The elastic-plastic removal grooves at the entrance of the polishing belts were deeper while the surface at the exit was flatter and smoother; this coincided with the mechanical calculation and nanoindentation experiment.

This study applied the experimental results of nanoindentation to the mechanical analysis of abrasives under different fixation methods and the removal and deformation analysis of single crystal SiC material. These research results are significant for the ultra-precision machining of single crystal SiC substrates.

Author Contributions: Conceptualization, J.P. and Q.Y.; Formal analysis, X.Z.; Funding acquisition, Q.Y.; Writing—original draft, J.P.; Writing—review & editing, W.L.

Funding: The authors would like to thank the NSFC-Guangdong Joint Fund Project (No. U1801259), the Guangdong Graduate Education Innovation Project (No. 2018JGXM35) and the Foshan Science and Technology Innovation Project for their financial support.

Conflicts of Interest: The authors declare no conflict of interest.

Nomenclature

Symbol	Nomenclature
f_s	Spindle feed(μm/s)
n_w	The rotational speed of workpiece (rpm)
n_s	The rotational speed of grinding wheel (rpm)
H	Hardness (Gpa)
E	Elasticity modulus (Gpa)
K_c	The fracture toughness (Mpa m$^{1/2}$)
ρ	The radius of curvature of grinding grains on the materials (mm)
F_0	The contact load (mN)
F_C	The contact load of C face (mN)
F_{Si}	The contact load of Si face (mN)
P^*	The critical load for brittle materials changing from plastic deformation to brittle fracture (mN)

P^*_C	The critical load for C face changing from plastic deformation to brittle fracture (mN)
φ	The proportion of magnetic particles in the MR fluids
$h_{\max}$	The maximum thickness of undeformed substrate (nm)
d_c	The critical grit cutting depth (nm)
R	The radius of the grinding wheels (mm)
L_w	The circumference of layers of grinding material of the cupped grinding wheels (mm)
C	The concentration of grinding materials of diamond grinding wheels
K_β	The coefficient relating to the shape of the grinding grains
d_w	The radius of grinding grain (μm)
r_m	The radius at any point on the surface of the workpieces
v_d	The volume fraction of diamonds in the grinding wheel
h_{wp}	The space between the workpieces and the lapping plates (mm)
F_{in}	The abrasive forces at the entrance of the polishing belts (μN)
S	The workpiece area (mm^2)
P_F	The polishing pressures (kPa)
P_m	The pressure produced by MR effects of the MR fluids (kPa)
P_d	The hydrodynamic pressure (kPa)
P_g	The buoyancy of MR fluids (kPa)
μ_0	The magnetic inductivity of a vacuum
μ	The magnetic inductivity of magnetic particles
η	The initial viscosity of the MR fluids (Pa.s)
v	Speed of polishing disk (m/s)
F_v	The volume fraction of the layers of material in the grinding wheels (g/cm^3)
P^*_{Si}	The critical load for Si face changing from plastic deformation to brittle fracture (mN)
K_b	The distribution coefficient of abrasives between the workpieces and lapping plates
H_m	The strength of the external magnetic fields (Gs)
h_0	The distance from the polishing plates to the surfaces of workpieces (mm)
t	The thickness of workpieces (mm)
Δ	The machining gap (mm)
W	The tooth width of layers of grinding material of the cupped grinding wheels (mm)
F_w	The average pressure on single abrasive (mN)
S_A	The actual contact area between a single abrasive and the workpieces (mm^2)
p	The lapping pressure (kPa)
d_g	The equivalent spherical diameter of diamond particles
f	The fraction of diamond particles that actively cut when grinding
$d_{\max}$	The maximum diameter of abrasives (mm)
F_{out}	The abrasive forces at the exit of the polishing belts (μN)

References

1. Yang, Y.; Duan, B.; Yuan, S.; Jia, H. Novel Developments and Challenges for the SiC Power Devices. Advanced Silicon Carbide Devices and Processing. In *Advanced Silicon Carbide Devices and Processing*; IntechOpen: London, UK, 2015; pp. 175–195.
2. Hamada, K.; Nagao, M.; Ajioka, M.; Kawai, F. SiC—Emerging Power Device Technology for Next-Generation Electrically Powered Environmentally Friendly Vehicles. *IEEE Trans. Electron. Devices* **2015**, *62*, 278–285. [CrossRef]
3. Nishio, J.; Ota, C.; Hatakeyama, T.; Shinohe, T.; Kojima, K.; Nishizawa, S.-I.; Ohashi, H. Ultralow-Loss SiC Floating Junction Schottky Barrier Diodes (Super-SBDs). *IEEE Trans. Electron. Devices* **2008**, *55*, 1954–1960. [CrossRef]
4. Neudeck, P.; Okojie, R.; Chen, L.-Y. High-temperature electronics—a role for wide bandgap semiconductors? *Proc. IEEE* **2002**, *90*, 1065–1076. [CrossRef]
5. Li, C.H.; Jernigan, G.G.; Jonker, B.T.; Goswami, R.; Hellberg, C.S. Solid phase epitaxy of 3C-SiC on Si (001). US Patent 20,160,118,465, 2016.

6. Pan, G.; Zhou, Y.; Luo, G.; Shi, X.; Zou, C.; Gong, H. Chemical mechanical polishing (CMP) of on-axis Si-face 6H-SiC wafer for obtaining atomically flat defect-free surface. *J. Mater. Sci.* **2013**, *24*, 5040–5047. [CrossRef]
7. Tanaka, H.; Shimada, S. Damage-free machining of monocrystalline silicon carbide. *CIRP Ann.* **2013**, *62*, 55–58. [CrossRef]
8. Yin, L.; Vancoille, E.Y.; Ramesh, K.; Huang, H. Surface characterization of 6H-SiC (0001) substrates in indentation and abrasive machining. *Int. J. Mach. Tools Manuf.* **2004**, *44*, 607–615. [CrossRef]
9. Meng, B.; Zhang, F.; Li, Z. Deformation and removal characteristics in nanoscratching of 6H-SiC with Berkovich indenter. *Mater. Sci. Semicond. Process.* **2015**, *31*, 160–165. [CrossRef]
10. Goel, S.; Luo, X.; Comley, P.; Reuben, R.L.; Cox, A. Brittle–ductile transition during diamond turning of single crystal silicon carbide. *Int. J. Mach. Tools Manuf.* **2013**, *65*, 15–21. [CrossRef]
11. Yan, J.; Gai, X.; Harada, H. Subsurface Damage of Single Crystalline Silicon Carbide in Nanoindentation Tests. *J. Nanosci. Nanotechnol.* **2010**, *10*, 7808–7811. [CrossRef] [PubMed]
12. Xiao, G.; To, S.; Zhang, G. The mechanism of ductile deformation in ductile regime machining of 6H SiC. *Comput. Mater. Sci.* **2015**, *98*, 178–188. [CrossRef]
13. Lee, H.; Kim, D.; An, J.; Lee, H.; Kim, K.; Jeong, H. Hybrid polishing mechanism of single crystal SiC using mixed abrasive slurry (MAS). *CIRP Ann.* **2010**, *59*, 333–336. [CrossRef]
14. Xiao, G.; Ren, M.; To, S. A Study of Mechanics in Brittle–Ductile Cutting Mode Transition. *Micromachines* **2018**, *9*, 49. [CrossRef]
15. Li, C.; Zhang, F.; Meng, B.; Ma, Z. Simulation and Experiment on Surface Morphology and Mechanical Properties Response in Nano-Indentation of 6H-SiC. *J. Mater. Eng. Perform.* **2017**, *26*, 1000–1009. [CrossRef]
16. Nawaz, A.; Mao, W.; Lu, C.; Shen, Y. Mechanical properties, stress distributions and nanoscale deformation mechanisms in single crystal 6H-SiC by nanoindentation. *J. Alloy. Compd.* **2017**, *708*, 1046–1053. [CrossRef]
17. Pang, K.H.; Tymicki, E.; Roy, A. Indentation in single-crystal 6H silicon carbide: Experimental investigations and finite element analysis. *Int. J. Mech. Sci.* **2018**, *144*, 858–864. [CrossRef]
18. Lu, J.; Luo, Q.; Xu, X.; Huang, H.; Jiang, F. Removal mechanism of 4H-and 6H-SiC substrates (0001 and 0001 over bar) in mechanical planarization machining. *Proc. Inst. Mech. Eng. B* **2019**, *233*, 69–76. [CrossRef]
19. Chen, Y. Study on Mechanism and Key Techniques of Ion Implantation Surface Modification Method for Ultra-Precision Machining. PhD Thesis, University of Queensland, Brisbane, Australia, 2015.
20. Mehregany, M.; Tong, L.; Matus, L.; Larkin, D. Internal stress and elastic modulus measurements on micromachined 3C-SiC thin films. *IEEE Trans. Electron. Devices* **1997**, *44*, 74–79. [CrossRef]
21. Stevanovic, M. Modeling thermal contact resistance between elastic hemisphere and elastic layer bonded to elastic substrate. In Proceedings of the ITherm 2002. Eighth Intersociety Conference on Thermal and Thermomechanical Phenomena in Electronic Systems (Cat. No.02CH37258), San Diego, CA, USA, 30 May–1 June 2002; IEEE: Piscataway, NJ, USA, 2002.
22. Ellyin, F. *Fatigue Damage, Crack Growth and Life Prediction*; Springer Science & Business Media: Berlin, Germany, 2012.
23. Broek, D. *Elementary Engineering Fracture Mechanics*; Springer Science & Business Media: Berlin, Germany, 2012.
24. Bradt, R.C.; Hasselman, D.P.H.; Munz, D.; Sakai, M.; Shevchenko, V.Y. (Eds.) *Fracture Mechanics of Ceramics: Composites, R-Curve Behavior, and Fatigue*; Springer Science & Business Media: Berlin, Germany, 2012.
25. Arif, M.; Xinquan, Z.; Rahman, M.; Kumar, S. A predictive model of the critical undeformed chip thickness for ductile–brittle transition in nano-machining of brittle materials. *Int. J. Mach. Tools Manuf.* **2013**, *64*, 114–122. [CrossRef]
26. Bifano, T.G.; Dow, T.A.; Scattergood, R.O. Ductile-regime grinding: a new technology for machining brittle materials. *J. Eng. Ind.* **1991**, *113*, 184–189. [CrossRef]
27. Shang, G. *Fundamental Research on Silicon Wafer Thinning by Ultra-precision Grinding*; Dalian University of Technology: Dalian, China, 2013.
28. Agarwal, S.; Rao, P.V. Experimental investigation of surface/subsurface damage formation and material removal mechanisms in SiC grinding. *Int. J. Mach. Tools Manuf.* **2008**, *48*, 698–710. [CrossRef]
29. Ioan, D.M.; Eckart, U. *Handbook of Lapping and Polishing (Manufacturing Engineering and Materials Processing)*; CRC Press: Boca Raton, FL, USA, 2006.
30. Czado, C.; Schmidt, T. *Mathematische Statistik*; Springer: Berlin/Heidelberg, Germany, 2011.

31. Evans, C.J.; Parks, R.E.; Roderick, D.J.; McGlauflin, M.L. Rapidly Renewable Lap: Theory and Practice. *CIRP Ann.* **1998**, *47*, 239–244. [CrossRef]
32. Pan, J.; Yan, Q. Material removal mechanism of cluster magnetorheological effect in plane polishing. *Int. J. Adv. Manuf. Technol.* **2015**, *81*, 2017–2026. [CrossRef]
33. Basim, G.B.; Vakarelski, I.U.; Moudgil, B.M. Role of interaction forces in controlling the stability and polishing performance of CMP slurries. *J. Colloid Interface Sci.* **2003**, *263*, 506–515. [CrossRef]
34. Feng, Z. Foundation of mathematics model of magnetorheological finishing. *Opt. Tech.* **2000**, *26*, 190–192.
35. Yang, P. *Numerical Analysis of Fluid Lubrication*; National Defense Industry Press: Beijing, China, 1998.

micromachines

MDPI

Article

The Influence of AlN Intermediate Layer on the Structural and Chemical Properties of SiC Thin Films Produced by High-Power Impulse Magnetron Sputtering

Nierlly Galvão [1,*], Marciel Guerino [1], Tiago Campos [1], Korneli Grigorov [2], Mariana Fraga [3,*], Bruno Rodrigues [1,4], Rodrigo Pessoa [1,4], Julien Camus [5], Mohammed Djouadi [5] and Homero Maciel [1,4]

1 Centro de Ciência e Tecnologia de Plasmas e Materiais—PlasMat, Instituto Tecnológico de Aeronáutica, São José dos Campos 12228-900, SP, Brazil; marcielguerino@yahoo.com.br (M.G.); moreiratiago22@gmail.com (T.C.); bruno.manzolli@gmail.com (B.R.); rspessoa@ita.br (R.P.); homero@ita.br (H.M.)

2 Space Research and Technology Institute, Acad. G. Bonchev Str. Bl.1, 1113 Sofia, Bulgaria; kgrigoro@abv.bg

3 Instituto de Ciência e Tecnologia, Universidade Federal de São Paulo, São José dos Campos 12231-280, SP, Brazil

4 Instituto Científico e Tecnológico, Universidade Brasil, Rua Carolina Fonseca 235, São Paulo 08230-030, Brazil

5 Institut des Matériaux Jean Rouxel IMN, UMR 6502, Université de Nantes, 2 rue de La Houssinière, BP 32229, 44322 Nantes CEDEX, France; jcamus@gmail.com (J.C.); abdou.djouadi@cnrs-imn.fr (M.D.)

* Correspondence: nierlly@gmail.com (N.G.); mafraga@ieee.org (M.F.)

Received: 10 February 2019; Accepted: 15 March 2019; Published: 22 March 2019

Abstract: Many strategies have been developed for the synthesis of silicon carbide (SiC) thin films on silicon (Si) substrates by plasma-based deposition techniques, especially plasma enhanced chemical vapor deposition (PECVD) and magnetron sputtering, due to the importance of these materials for microelectronics and related fields. A drawback is the large lattice mismatch between SiC and Si. The insertion of an aluminum nitride (AlN) intermediate layer between them has been shown useful to overcome this problem. Herein, the high-power impulse magnetron sputtering (HiPIMS) technique was used to grow SiC thin films on AlN/Si substrates. Furthermore, SiC films were also grown on Si substrates. A comparison of the structural and chemical properties of SiC thin films grown on the two types of substrate allowed us to evaluate the influence of the AlN layer on such properties. The chemical composition and stoichiometry of the samples were investigated by Rutherford backscattering spectrometry (RBS) and Raman spectroscopy, while the crystallinity was characterized by grazing incidence X-ray diffraction (GIXRD). Our set of results evidenced the versatility of the HiPIMS technique to produce polycrystalline SiC thin films at near-room temperature by only varying the discharge power. In addition, this study opens up a feasible route for the deposition of crystalline SiC films with good structural quality using an AlN intermediate layer.

Keywords: high-power impulse magnetron sputtering (HiPIMS); silicon carbide; aluminum nitride; thin film; Rutherford backscattering spectrometry (RBS); grazing incidence X-ray diffraction (GIXRD); Raman spectroscopy

1. Introduction

Silicon carbide (SiC) has been proven to be a promising material for microelectronic applications due to its excellent physical and electronic properties, such as high surface hardness, wide bandgap, and high thermal conductivity at low and high temperatures [1–6]. These outstanding properties make it an attractive material for the development of harsh-environment devices such as Micro-Electro-Mechanical Systems (MEMS) and power electronics [1,2,7–9]. In a recent article, Dinh et al. showed an on-chip SiC MEMS device for efficient thermal management [10]. Furthermore, the strain effect in a highly doped 3C–SiC-on-glass substrate for mechanical sensors was also recently reported by Phan et al. [11]. The 3C–SiC bridges were also investigated under the consideration of Joule heating [12].

For microelectronic device applications, it is desirable for SiC thin films to be grown on Si substrates because their manufacturing processes are based on Si microfabrication technology, which is compatible with standard industrial processes [8,13–15]. It is difficult to grow high-quality crystalline SiC (c-SiC) films on Si substrates at low temperatures (<300 °C) due to a large mismatch between their lattice constant (about 20%) and thermal expansion coefficients (about 8%), which usually affects the final properties of the grown material [16]. In order to reduce these effects, an intermediate or buffer layer may be added. For this purpose, aluminum nitride (AlN) thin film is frequently used since it presents minimum mismatching in the lattice constant (less 1%) with SiC, and has a similar thermal expansion coefficient [17–20].

Meguro et al. investigated the formation of a SiC interfacial buffer layer on AlN/Si substrates at a low temperature by low-pressure chemical vapor deposition (LPCVD) [17]. Nakazawa et al. reported the epitaxial growth of SiC films on an AlN layer on Si (100) substrates by ultralow-pressure chemical vapor deposition. Jeong et al. investigated the Raman scattering characteristics of 3C–SiC films deposited on AlN/Si substrates using the atmosphere pressure chemical vapor deposition (APCVD) technique [19]. Huang et al. demonstrated the formation of SiC quantum dots (SiC QDs) on AlN films using low-frequency inductively coupled plasma (LF-ICP)-assisted magnetron sputtering [20]. To our knowledge, the study here presented is the first to report the growth of high-power impulse magnetron sputtering (HiPIMS) of SiC films on AlN/Si substrates.

The achievement of good crystallinity in the SiC thin films is a desirable feature since it influences different material properties [21]. As the SiC thin films deposited at low temperature grow in amorphous or nanocrystalline structures, post-treatment such as annealing, to improve the material crystallinity, is necessary. Although there are several well-known techniques for synthesizing SiC thin films, their composition and final properties may vary considerably with the applied method [3,19]. Low-pressure plasma-based techniques have been extensively investigated, particularly those that allow the deposition at near-room temperatures, such as plasma-enhanced chemical vapor deposition (PECVD) and magnetron sputtering [1,3,22–25]. Along with the magnetron sputtering derivations, the high-power impulse magnetron sputtering (HiPIMS) technique appears to be very attractive due to its ability to generate high-density plasmas and a high degree of ionization of the sputtered atoms [26–31]. These properties allow sufficient energy for the rearrangement of atoms/molecules during the growth of the film, thus facilitating the formation of crystalline phases. Some reports have demonstrated that, depending on the deposition parameters and target composition, around 50–90% of the sputtering atoms are in an ionized state [24,28,30]. This occurs because of the mechanism in which the HiPIMS power supply applies the power over the magnetron target for generating the plasma, namely high-power pulses, low frequency, and low duty cycles (lower than 10%) [28,30–32]. Interesting reviews on HiPIMS were written by Sarakinos et al. [30] and Gudmundsson et al. [31].

Although the HiPIMS source is applied in the synthesis of various metals and semiconductor materials, there is a clear lack of studies focusing on the growth of SiC thin films using this technique. The studies related to this topic are focused on Ti–Si–C and SiCN films using HiPIMS [26,33]. In the work of Alami et al. [33], the effect of processing parameters such as gas pressure, substrate geometry, and distance of the target substrate on some properties of the as-deposited Ti–Si–C film was investigated. They observed that the Ti–Si–C film quality could be improved by the HiPIMS technique [33]. Pusch et al. performed a comparison between SiCN films deposited with different target configurations and techniques, i.e., radiofrequency (RF), direct current (DC), and HiPIMS [26]. Leal et al. deposited SiC thin films on Si substrates by HiPIMS using a SiC target [34]; however, only amorphous films were obtained. In this article, we explore the structural and chemical properties of polycrystalline SiC films grown at room temperature on Si and AlN/Si substrates by the HiPIMS technique. The composition, chemical bonding, structure, and crystallinity of the samples were investigated by Rutherford backscattering spectrometry (RBS), Raman spectroscopy, and grazing incidence X-ray diffraction (GIXRD).

2. Materials and Methods

2.1. Deposition Method

SiC thin films were grown onto polished p-type Si (100) wafers, either with or without an AlN layer, via HiPIMS in a high-vacuum chamber with a background pressure of 6×10^{-6} Torr. More details about the HiPIMS reactor can be found elsewhere [34,35]. The working pressure of the argon gas (99.999%, White-Martins, São José dos Campos, Brazil) was maintained at 3 mTorr for a corresponding flow rate of 20 sccm (standard centimeter cubic per minute). The target was a commercial high-purity SiC (99.5%, Kurt J. Lesker Company, Jefferson Hills, PA, USA) with a diameter of 4 inches. For film growth, the applied power from the HiPIMS power supply (HIP^3 5 kW, Solvix SA, Villaz-Saint-Pierre, Switzerland) was 200 W and 400 W. In all cases, the duty cycle was fixed at 5%, frequency at 500 Hz, and pulse time at 100 μs. To remove the target surface contaminants, a pre-sputtering time of 200 W for 10 min was applied. In addition, the substrate holder was maintained at a floating potential, whereas the deposition time and the target-substrate distance were fixed at 60 min and 60 mm, respectively.

The AlN/Si substrates were provided by the "Institut des Matériaux Jean Rouxel in Nantes University". In these substrates, the sputtered AlN intermediate layer had a thickness of 1300 nm and a (002) crystallographic orientation. For more details, see References [36,37].

2.2. Characterization Techniques

Rutherford backscattering spectrometry (RBS) was used to measure the elemental composition, stoichiometry, and the thickness of the as-deposited SiC thin films. The measurements were performed with a pelletron accelerator using 2 MeV $4He^+$ beam with a particle detector positioned at 170° from the incident beam. The RBS spectra were analyzed using the computer code RUMP (RBS analysis package) developed by L. R. Doolittle from Cornell University [38]. To verify the accuracy of the RBS thickness measurements and the thickness uniformity, mechanical profilometry (P-7 Stylus Profiler, KLA Tencor, Milpitas, CA, USA) measurements were performed.

The crystallinity of the SiC films was inferred from GIXRD with incidence angles (ω) of 1.0°, 1.5°, and 2.0° using an X-ray diffractometer (PW1830/1840, Philips, Amsterdam, The Netherlands) with CuK_α radiation. For Raman spectroscopy measurements, a model 2000 Renishaw system (Renishaw, Wotton-under-Edge, UK), equipped with an Ar ion laser (514.5 nm) was used. Raman spectra were obtained at room temperature in the range of 400–1800 cm^{-1}.

3. Results and Discussion

3.1. Chemical Composition and Stoichiometry

Figure 1 shows the experimental and simulated RBS spectra of the as-deposited SiC thin films on Si and AlN/Si substrates for the 200 W and 400 W conditions. Table 1 summarizes the results of the RBS spectra analysis.

Figure 1a depicts the spectrum of SiC deposited on Si at 200 W and the simulation reveals a film with a total thickness of approximately 1200 nm, with a highly non-homogenous elemental distribution throughout the film depth. To better visualize the variation of the stoichiometry throughout the film depth, the simulation comprised five sublayers which are identified in Table 1. The top layer's stoichiometry comprised 260 nm of pure SiC with less than 13% oxygen. The middle part (approximately 400 nm) consisted of SiC with 10% excess carbon. The next layer with a thickness of 250 nm had about 50% carbon excess and a substantial drop in the oxygen content was observed. The subsequent layer of 170 nm was fully stoichiometric, followed by the last layer of 145 nm adjacent to the Si surface, where 10% carbon excess was found. When investigating the incorporated oxygen in the first layers of the SiC film, Medeiros et al. observed the unintentional doping of SiC_xN_y thin films by oxygen contamination coming from the vacuum environment of the magnetron co-sputtering system [35]. In this work, RBS results showed that all samples contained significant amounts of oxygen (up to 16%). Further, X-ray photoelectron spectroscopy (XPS) results showed that most of this oxygen is located in the film surface [35]. These results corroborate with the RBS analysis presented in Figure 1a. In addition, Pomaska et al. presented studies on the unintentional doping by oxygen contamination where they demonstrated that the oxygen incorporation was influenced the microstructural, electronic, and optical properties of the SiC films [39]. It has been shown that oxygen incorporation during film deposition increases the crystallinity of SiC films, consistent with findings observed in this work.

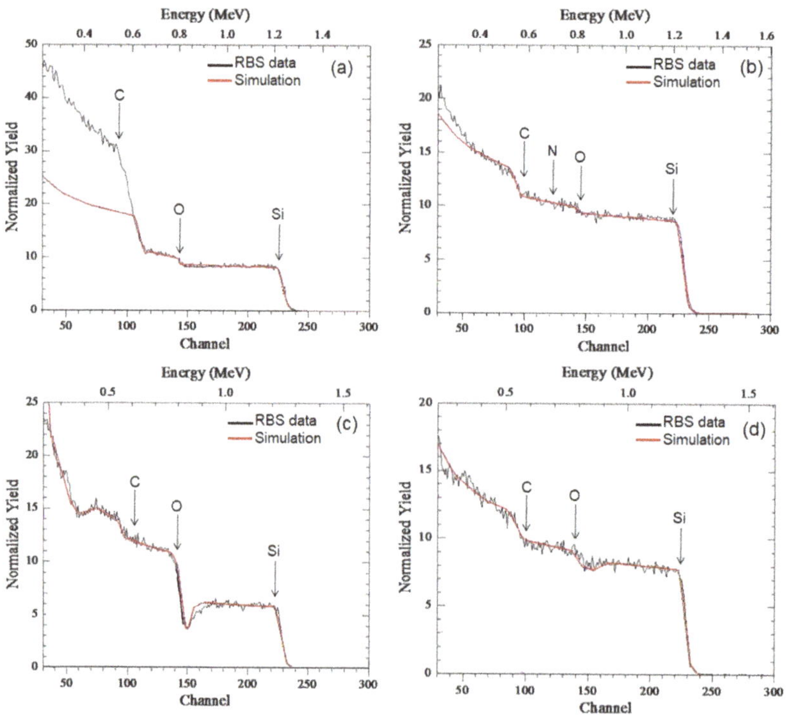

Figure 1. Experimental and simulated Rutherford backscattering spectrometry (RBS) spectra of the SiC films deposited on (**a**) Si substrate at 200 W; (**b**) Si substrate at 400 W; (**c**) AlN/Si substrate at 200 W and; (**d**) AlN/Si substrate at 400 W.

For the SiC grown on the Si substrate at 400 W (Figure 1b), the analysis of the RBS spectra indicated that the total film thickness was around 1500 nm. The film exhibited a pure and stoichiometric composition of SiC throughout the entire depth, although two zones could be distinguished as presented in Table 1. Beyond the SiC, SiO_2, and SiN phases, there were O and N contaminants. This sputtering condition resulted in a heterogeneous film composition with variable elemental depth distributions. In general, the higher power deposition energy, as in this case, leads to Ar ions striking onto the film surface with high energy, which contributes to the formation of chemical phases.

Of course, if different impurities act as film constituents, they are involved in the film composition forming stable bonds (SiO_2; SiN).

From the thickness results of the SiC films grown at 200 W (1200 nm) and 400 W (1500 nm), it is possible to observe that although the applied power is twice as high, there was a small increase in the deposition rate for SiC films on the Si substrate. In conventional sputtering processes, the deposition rate of the SiC film increases linearly with the sputtering power [23,24]. In general, HiPIMS exhibits different growth mechanisms and lower deposition rates than those observed for conventional sputtering processes [30,31,40]. Different effects have been considered to explain the differences between DC and HiPIMS deposition rates. There are three main reasons considered [41]: (i) the less-than-linear increase of the sputtering yield with increasing ion energy, ion return to the target, and self-sputtering; (ii) ion return to the target and self-sputtering; and (iii) changes due to greater film density, limited sticking, and self-sputtering on the substrate.

Table 1. Results of the RBS analysis.

Sample	No. of Layers	Composition by Layer [1]	Layer Thickness
SiC/Si 200 W	5	1. SiC stoichiometry with less than 13% oxygen. 2. SiC with about 10% excess carbon. 3. SiC with about 50% excess carbon. 4. SiC stoichiometry. 5. SiC with about 10% excess carbon.	1. 260 nm 2. 400 nm 3. 250 nm 4. 170 nm 5. 145 nm
SiC/Si 400 W	2	1. SiC—86%; SiO_2 phase—4% dispersed in that volume; O—5%; N—5%. 2. SiC—50% and SiN—50%.	1. ~900 nm 2. ~600 nm
SiC/AlN/Si 200 W	2	1. SiC stoichiometry—56%; C solid state and O contamination in 44%. 2. AlN layer stoichiometry.	1. ~930 nm 2. ~1300 nm
SiC/AlN/Si 400 W	2	1. SiC stoichiometry—80% with 20% C solid state and O contamination in volume. 2. AlN layer stoichiometry.	1. ~1360 nm 2. ~1300 nm

[1] Layer 1 refers to the layer at the top of the film.

For the SiC film deposited on AlN/Si at 200 W, the total film thickness was around 930 nm (Figure 1c). The film composition was rather homogenous and consisted of 56% pure SiC, while the remaining 44% of the film was composed of C and O in the bulk of the film. The intermediate layer of AlN consisted of 1300 nm thick sub-stoichiometric AlN with 5% less nitrogen, resulting in some point defects. Note that in this case the substrate change provided the growth of a high stoichiometric SiC film. Relative to film thickness, it is evidenced from data presented in Table 1 that the change of Si with AlN/Si substrate promoted the decrease in the thickness of the SiC film. Although sputtering processes have deposition rates that are independent of the substrate type, the film nucleation process and consequent crystallization and compaction are dependent. Nivedita et al. confirmed some of these observations when depositing RF-sputtered Fe–Ga thin films on MgO, quartz, and Si substrates [42]. Indeed, the next topic shows that the crystallization of SiC is improved for films deposited on Si. Crystalline films tend to have greater roughness and even porosity in comparison with amorphous films, which consequently increases the final thickness [43].

The SiC thin film deposited on AlN/Si at 400 W (Figure 1d) exhibited a high percentage of purely stoichiometric SiC film, with the presence of C and O in volume. However, for this condition the estimate of the thickness by RBS was limited due to the loss of the energy via scattering albeit within certain limits, e.g., above channel n° 90 (Figure 1d), the thickness could be estimated as being around 1360 nm. Ultimately, the elemental depth distribution throughout the film thickness was uniform, which made the present method and processing conditions very useful for the achievement of high-quality SiC thin film deposition.

Finally, from the results in Table 2, it was possible to observe that the calculated deposition rates of the SiC films were in agreement with the profilometry measurements. With regard to the deposition rates measured by profilometry, and where the film thicknesses were measured at different points during the formation of the film, it was possible to evaluate the uniformity of the film thickness, which exhibited a 3% variation throughout the substrate. In fact, the greater the target–substrate distance in processes performed by magnetron sputtering, the better the uniformity of the film formed, where a distance of 60 mm was used. With regard to the film morphology, in previous work [44] Atomic force microscopy (AFM) analyses of the SiC/AlN/Si film and the AlN/Si film were performed, showing films with rough surfaces and with grain sizes smaller than 100 nm.

Table 2. Deposition rate of the SiC films.

Sample	Power (W)	Deposition Rate—RBS (nm/min)	Deposition Rate—Profilometer (nm/min)
SiC/Si	200	20.0	14.0 ± 0.3
SiC/Si	400	25.0	19.6 ± 0.4
SiC/AlN/Si	200	15.5	12.5 ± 0.4
SiC/AlN/Si	400	22.7	24.0 ± 0.5

3.2. Structural Analysis

Figures 2–4 show the patterns of grazing incidence angles of 1.0°, 1.5°, and 2.0°, respectively. The Bragg reflections suggest the existence of α and β SiC nanocrystalline structures. Although the patterns exhibit the SiC phase, it is not possible to determine which of the SiC phases are present because some diffraction peaks of α and β SiC might overlap [45]. The carbon phase at ~25° was also visible and confirmed the RBS results indicating an excess of C (Table 1). Lastly, two broad peaks at ~47° and 55° were assigned as unidentified. While some studies have attributed these peaks to the SiC polymorph phase [46,47], others often define them as being C or Si phases [48–50]. When comparing the results from the GIXRD with an incidence angle of 2° (Figure 4) with the smaller angle results, the variation of the crystalline phases with the depth of the film was clearly noted.

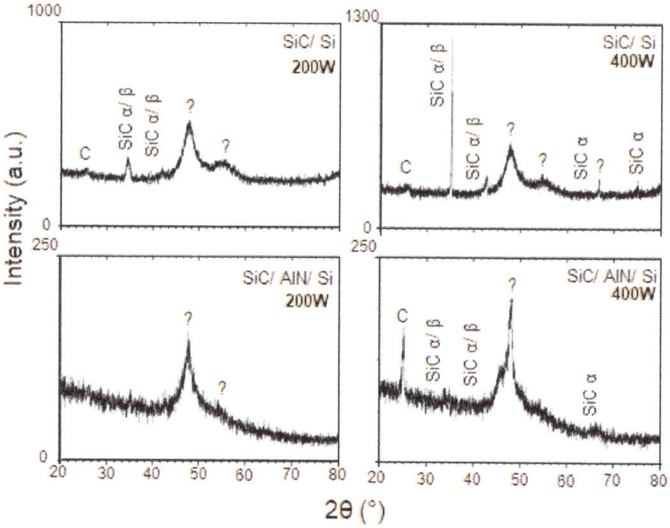

Figure 2. Grazing incidence X-ray diffraction (GIXRD) patterns of the SiC thin films at a grazing angle of 1.0°.

In addition, Figure 2 suggests the existence of SiC nanocrystalline structures achieved without substrate heating. By varying the GIXRD incidence angle, the resulting film could be analyzed in depth. A comparison of the patterns between SiC/Si and SiC/AlN/Si for both angles of incidence (1° and 1.5°), revealed that the phase observed at approximately 36°, using the 1.5° GIXRD as the incidence angle, no longer existed in the pattern obtained with the smallest angle (1.0°). This result pointed to the existence of phase and crystallinity variations with depth.

The film deposited on the AlN layer showed dislocation of the SiC peak between 35 and 36° in the GIXRD patterns. This dislocation may be attributed to the following reasons: (i) film stress; (ii) interference of the substrate (SiC/Si and SiC/AlN interface); or (iii) residual stress. More studies are necessary to better understand this observation.

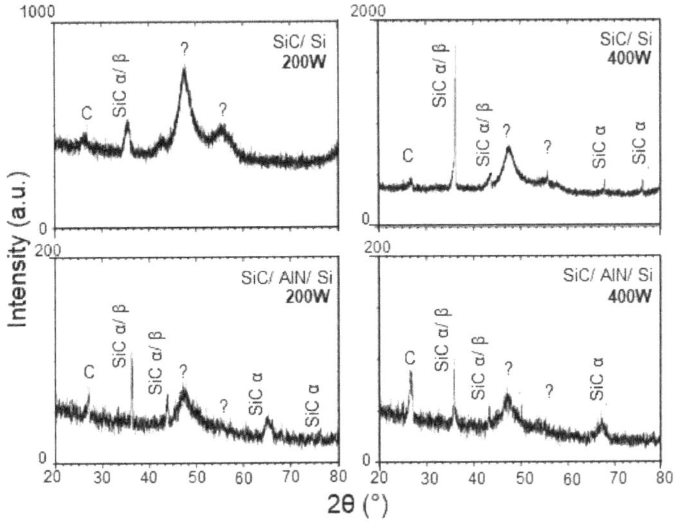

Figure 3. GIXRD patterns of the SiC thin films at a grazing angle of 1.5°.

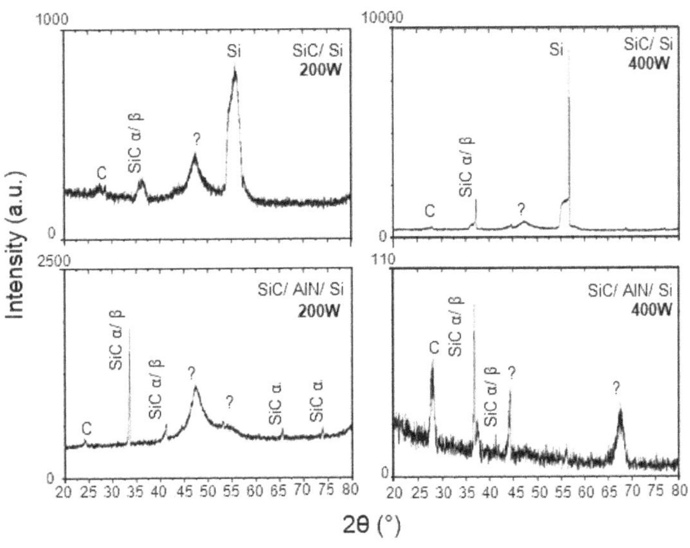

Figure 4. GIXRD patterns of the SiC thin films at a grazing angle of 2°.

3.3. Raman Spectroscopy

Figure 5 shows the results of Raman spectroscopy used to identify the bonds present in the films. The Raman spectra for the SiC films deposited at 200 W in both substrates (with and without the AlN layer) are presented in Figure 5a, showing a very visible and well-defined Si peak at 519.41 cm^{-1}. Since the difference in thickness between both films was small, the substrate had an accentuated influence. In addition to Si, the SiC film deposited on the AlN layer showed (i) a peak relative to AlN at ~652.20 cm^{-1}; (ii) peaks for SiC and Si in the regions between 741–894 cm^{-1} and 906–1109 cm^{-1}, respectively; (iii) and a broad carbon band at 1370–1625 cm^{-1}. Except for the AlN peak, the SiC/Si spectrum exhibited signals at similar regions to that of the SiC/AlN/Si spectrum. However, the regions relative to SiC and Si were more visible, and the C band region had a more explicit separation in two peaks (D and G bands), but with a low definition of the disorder band. The D band was attributed to the disorder or polycrystalline carbon and the G band to the graphite-like carbon [19,50].

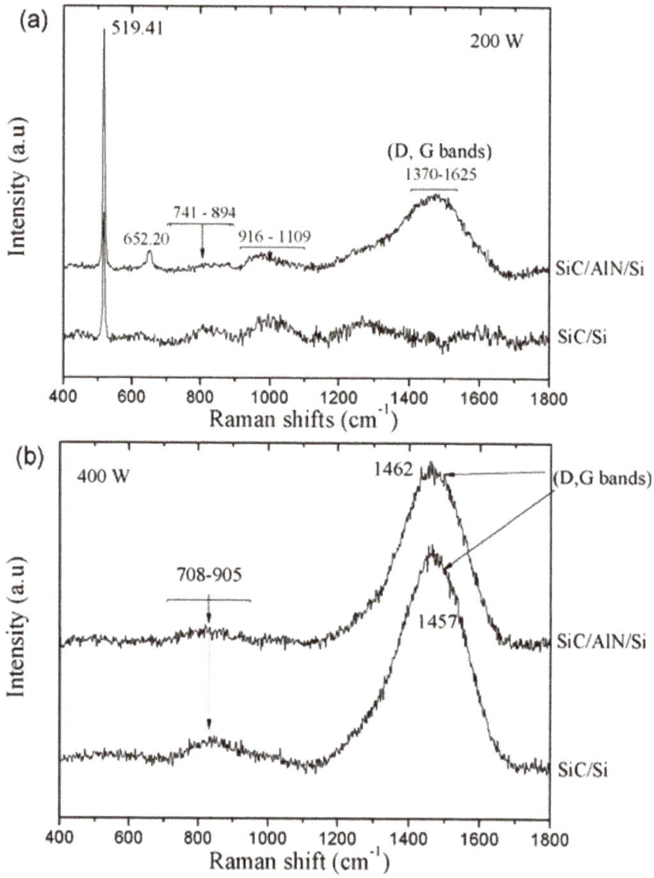

Figure 5. Raman spectra of SiC thin films on both substrates: (**a**) as-deposited at 200 W and (**b**) as-deposited at 400 W.

For 400 W (Figure 5b), both substrates exhibited similar behavior following the deposition of SiC films. In these spectra, the substrate signal was not detected and only peaks corresponding to carbon at 1457 and 1462 cm^{-1} were evident. With the exception of the spectrum of the SiC film at 200 W of SiC/Si, where the peak of the C band was not overlapped, all spectra of the "as-deposited" SiC film showed overlapping C, D, and G bands between ~1457 and 1462 cm^{-1}. Ferrari and Robertson [51] reported that for C–C bands, the Raman spectrum was influenced by factors such as disorder, clustering of the sp^2 phase, the presence of sp^2 C-rings or C-chains, and the ratio of sp^2 to sp^3 (I(D)/I(G)). Thus, with an increasing disorder of the C phase, the G peak position can be moved, and the D and G peaks will therefore overlap [19,50,51].

4. Conclusions

The influence of an AlN intermediate layer on the structural and chemical properties of HiPIMS SiC films grown on Si substrates was investigated using RBS, Raman spectroscopy, and GIXRD. The effect of the applied power (200 W and 400 W) was also considered. It was observed that the HiPIMS of SiC films can exhibit a complex growth mechanism and, depending on the process parameters, leads to the formation of films with an inhomogeneous composition throughout the depth of the Si substrate and a homogeneous composition for the AlN/Si substrate. This was verified by GIXRD using three different incidence angles (1.0°, 1.5°, and 2.0°) which, besides confirming the RBS results, also evidenced the variation of crystallinity with the depth of the film. Raman spectroscopy analysis indicated the presence of Si–C bonds and that the C–C bond region was separated into two peaks (D and G bands), but with a low definition of the disorder band. In summary, the results demonstrated that the HiPIMS technique and the use of an AlN intermediate layer allowed for the deposition of crystalline SiC films of good quality, without the need for substrate heating, with approximately 1.5 μm (at 400 W) in only 60 min, i.e., at a deposition rate of 25 nm/min. The good chemical and physical properties of the HiPIMS SiC films deposited on AlN/Si substrates highlights their potential benefits in nanotechnological applications. Indeed, we recently proposed the thermal decomposition of SiC thin films using a CO_2 laser beam without a vacuum chamber for graphene synthesis. The use of an AlN layer proved to be important because it reduces the thermal stress between SiC and Si materials [44]. Other applications will be the subject of further work.

Author Contributions: N.G., M.G., T.C., K.G., R.P., and J.C. conducted the investigation and methodology, N.G., K.G., M.F., B.R., R.P., M.D., and H.M. participated in the writing of the original draft.

Funding: The following funding sources including the Brazilian agency program FAPESP/MCT/CNPq-PRONEX (grant No. 11/50773-0), MCTI/CNPq/Universal (grants No. 459688/2014-6, 437921/2018-2, and 421317/2018-3), FAPESP (grant No. 18/01265-1 and 14/18139-8), CAPES/ITA (process No. 23038.005802/2014-98), CAPES/PVE (process No. 88881.064970/2014-01 and BEX9796/12-6), PNPD-CAPES, and CNPq (grants No. 305496/2012-3 and 446545/2014-7) are gratefully acknowledged for their support.

Acknowledgments: We would like to thank LAS-INPE for the Raman spectroscopy and GIXRD measurements, and LAMFI-USP for the RBS measurements. The authors also thank Dr Julio César Sagas from UDESC-Santa Catarina/Brazil for discussions and operation of the HiPIMS reactor.

Conflicts of Interest: The authors declare no conflict of interest.

References

1. Fraga, M.A.; Bosi, M.; Negri, M. Silicon Carbide in Microsystem Technology—Thin Film Versus Bulk Material. In *Advanced Silicon Carbide Devices and Processing*, 1st ed.; Saddow, S., la Via, F., Eds.; InTech: Rijeka, Croatia, 2015; pp. 1–31.
2. Fraga, M.A.; Furlan, H.; Pessoa, R.S.; Massi, M. Wide bandgap semiconductor thin films for piezoelectric and piezoresistive MEMS sensors applied at high temperatures: An overview. *Microsyst. Technol.* **2014**, *20*, 9–21. [CrossRef]
3. Pessoa, R.S.; Fraga, M.A.; Santos, L.V.; Massi, M.; Maciel, H.S. Nanostructured thin films based on TiO_2 and/or SiC for use in photoelectrochemical cells: A review of the material characteristics, synthesis and recent applications. *Mater. Sci. Semicond. Process.* **2015**, *29*, 56–68. [CrossRef]
4. Maboudian, R.; Carraro, C.; Senesky, D.G.; Roper, C.S. Advances in silicon carbide science and technology at the micro- and nanoscales. *J. Vac. Sci. Technol. A* **2013**, *31*, 050805. [CrossRef]
5. Müller, G.; Krötz, G.; Niemann, E. SiC for sensors and high-temperature electronics. *Sens. Actuators A Phys.* **1994**, *43*, 259–268. [CrossRef]

6. Huang, J.; Wang, L.; Wen, J.; Wang, Y.; Lin, C.; Ostling, M. Effect of annealing on SiC thin films prepared by pulsed laser deposition. *Diam. Relat. Mater.* **1999**, *8*, 2099–2102. [CrossRef]
7. Ledermann, N.; Baborowski, J.; Muralt, P.; Xantopoulos, N.; Tellenbach, J. Sputtered silicon carbide thin films as protective coating for MEMS applications. *Surf. Coat. Technol.* **2000**, *125*, 246–250. [CrossRef]
8. Sarro, P.M. Silicon carbide as a new MEMS technology. *Sens. Actuators A Phys.* **2000**, *82*, 210–218. [CrossRef]
9. Fraga, M.A.; Pessoa, R.S.; Massi, M.; Maciel, H.S. Applications of SiC-Based Thin Films in Electronic and MEMS Devices. In *Physics and Technology of Silicon Carbide Devices*, 1st ed.; Hijikata, Y., Ed.; InTech: Rijeka, Croatia, 2012; pp. 313–336.
10. Dinh, T.; Phan, H.P.; Kashaninejad, N.; Nguyen, T.K.; Dao, D.V.; Nguyen, N.T. An On-Chip SiC MEMS Device with Integrated Heating, Sensing, and Microfluidic Cooling Systems. *Adv. Mater. Interfaces* **2018**, *5*, 1800764. [CrossRef]
11. Phan, H.P.; Nguyen, T.K.; Dinh, T.; Cheng, H.H.; Mu, F.; Lacopi, A.; Hold, L.; Dao, D.V.; Suga, T.; Senesky, D.G.; et al. Strain Effect in Highly-Doped n-Type 3C-SiC-on-Glass Substrate for Mechanical Sensors and Mobility Enhancement. *Phys. Status Solidi A* **2018**, *215*, 1870054. [CrossRef]
12. Balakrishnan, V.; Dinh, T.; Phan, H.P.; Kozeki, T.; Namazu, T.; Dao, D.V.; Nguyen, N.T. Steady-state analytical model of suspended p-type 3C–SiC bridges under consideration of Joule heating. *J. Micromech. Microeng.* **2017**, *27*, 075008. [CrossRef]
13. Roumié, M.; Tabbal, M.; Nsouli, B.; Said, A. Determination of stoichiometry in silicon carbide materials using elastic backscattering spectrometry. *Nucl. Instrum. Methods Phys. Res. B* **2007**, *260*, 637–641. [CrossRef]
14. Tabbal, M.; Said, A.; Hannoun, E.; Christidis, T. Amorphous to crystalline phase transition in pulsed laser deposited silicon carbide. *Appl. Surf. Sci.* **2007**, *253*, 7050–7059. [CrossRef]
15. Zhao, Y.M.; Sun, G.S.; Liu, X.S.; Li, J.Y.; Zhao, W.S.; Wang, L.; Li, J.M.; Zeng, Y.P. Heteroepitaxial Growth of 3C-SiC on Si (111) Substrate Using AlN as a Buffer Layer. *Mater. Sci. Forum* **2009**, *600–603*, 251–254. [CrossRef]
16. Severino, A. 3C-SiC epitaxial growth on large area silicon: Thin films. In *Silicon Carbide Epitaxy*, 1st ed.; la Via, F., Ed.; Research Singpost: Kerala, India, 2012; pp. 145–191.
17. Meguro, K.; Narita, T.; Noto, K.; Nakazawa, H. Formation of an interfacial buffer layer for 3C-SiC heteroepitaxy on AlN/Si substrates. *Mater. Sci. Forum* **2014**, *778–780*, 251–254. [CrossRef]
18. Nakazawa, H.; Suzuki, D.; Narita, T.; Meguro, K.; Tsuchiya, M. Growth of silicon carbide on Si (100) substrate with an intermediate aluminum nitride layer by ultralow-pressure chemical vapor deposition using monomethylsilane. *J. Cryst. Growth* **2015**, *418*, 52–56. [CrossRef]
19. Jeong, J.; Jang, K.; Lee, H.S.; Chung, G.; Kim, G. Raman scattering studies of polycrystalline 3C-SiC deposited on SiO_2 and AlN thin films. *Phys. B Condens. Matter* **2009**, *404*, 7–10. [CrossRef]
20. Huang, S.Y.; Xu, S.; Long, J.D.; Sun, Z.; Chen, T. Plasma-reactive SiC quantum dots on polycrystalline AlN films. *Phys. Plasmas* **2006**, *13*, 023506. [CrossRef]
21. Fraga, M.A.; Pessoa, R.S.; Maciel, H.S.; Massi, M. Recent Developments on Silicon Carbide Thin Films for Piezoresistive Sensors Applications. In *Silicon Carbide—Materials, Processing and Applications in Electronic Devices*; Mukherjee, M., Ed.; InTech: Rijeka, Croatia, 2011; pp. 369–388.
22. Tang, H.; Tan, S.; Huang, Z.; Dong, S.; Jiang, D. Surface morphology of α-SIC coatings deposited by RF magnetron sputtering. *Surf. Coat. Technol.* **2005**, *197*, 161–167. [CrossRef]
23. Gou, L.; Qi, C.; Ran, J.; Zheng, C. SiC film deposition by DC magnetron sputtering. *Thin Solid Film.* **1999**, *345*, 42–44. [CrossRef]
24. Medeiros, H.S.; Pessoa, R.S.; Maciel, H.S.; Massi, M.; Tezani, L.L.; Leal, G.; Galvão, N.K.A.M.; da Silva Sobrinho, A.S. Amorphous Silicon Carbide Thin Films Deposited by Magnetron Co-Sputtering: Effect of Applied Power and Deposition Pressure on Film Characteristics. *ECS Trans.* **2012**, *49*, 375–382. [CrossRef]
25. Lei, Y.M.; Yu, L.H.; Cheng, L.L.; Ren, C.X.; Zou, S.C.; Wong, S.P.; Chen, D.H.; Wilson, I.H. IR studies of reactive DC magnetron sputtered SiC films on silicon using effective medium theory. *Mater. Lett.* **2000**, *43*, 215–219. [CrossRef]
26. Pusch, C.; Hoche, H.; Berger, C.; Riedel, R.; Ionescu, E.; Klein, A. Influence of the PVD sputtering method on structural characteristics of SiCN-coatings—Comparison of RF, DC and HiPIMS sputtering and target configurations. *Surf. Coat. Technol.* **2011**, *205*, s119–s123. [CrossRef]
27. Bräuer, G.; Szyszka, B.; Vergo, M.; Bandorf, R. Magnetron sputtering—Milestones of 30 years. *Vacuum* **2010**, *84*, 1354–1359. [CrossRef]

28. Anders, A. High power impulse magnetron sputtering and related discharges: Scalable plasma sources for plasma-based ion implantation and deposition. *Surf. Coat. Technol.* **2010**, *204*, 2864–2868. [CrossRef]
29. Aijaz, A.; Lundin, D.; Larsson, P.; Helmersson, U. Dual-magnetron open field sputtering system for sideways deposition of thin films. *Surf. Coat. Technol.* **2010**, *204*, 2165–2169. [CrossRef]
30. Sarakinos, K.; Alami, J.; Konstantinidis, S. High power pulsed magnetron sputtering: A review on scientific and engineering state of the art. *Surf. Coat. Technol.* **2010**, *204*, 1661–1684. [CrossRef]
31. Gudmundsson, J.T. The high power impulse magnetron sputtering discharge as an ionized physical vapor deposition tool. *Vacuum* **2010**, *84*, 1360–1364. [CrossRef]
32. Alami, J.; Bolz, S.; Sarakinos, K. High power pulsed magnetron sputtering: Fundamentals and applications. *J. Alloys Compd.* **2009**, *483*, 530–534. [CrossRef]
33. Alami, J.; Eklund, P.; Emmerlich, J.; Wilhelmsson, O.; Jansson, U.; Högberg, H.; Hultman, L.; Helmersson, U. High-power impulse magnetron sputtering of Ti–Si–C thin films from a Ti_3SiC_2 compound target. *Thin Solid Film.* **2006**, *515*, 1731–1736. [CrossRef]
34. Leal, G.; Campos, T.M.B.; da Silva Sobrinho, A.S.; Pessoa, R.S.; Maciel, H.S.; Massi, M. Characterization of SiC thin films deposited by HiPIMS. *Mater. Res.* **2014**, *17*, 472–476. [CrossRef]
35. Medeiros, H.S.; Pessoa, R.S.; Sagas, J.C.; Fraga, M.A.; Santos, L.V.; Maciel, H.S.; Massi, M.; da Silva Sobrinho, A.S.; Costa, M.E.H.M. Effect of nitrogen content in amorphous $SiC_xN_yO_z$ thin films deposited by low temperature reactive magnetron co-sputtering technique. *Surf. Coat. Technol.* **2011**, *206*, 1787–1795. [CrossRef]
36. Belkerk, B.E.; Soussou, A.; Carette, M.; Djouadi, M.A.; Scudeller, Y. Structural-dependent thermal conductivity of aluminium nitride produced by reactive direct current magnetron sputtering. *Appl. Phys. Lett.* **2012**, *101*, 151908. [CrossRef]
37. Aissa, K.A.; Achour, A.; Camus, J.; Brizoual, L.L.; Jouan, P.; Djouadi, A. Comparison of the structural properties and residual stress of AlN films deposited by dc magnetron sputtering and high power impulse magnetron sputtering at different working pressures. *Thin Solid Film.* **2014**, *550*, 264–267. [CrossRef]
38. Doolittle, L.R. Algorithms for the rapid simulation of Rutherford backscattering spectra. *Nucl. Instrum. Methods Phys. Res. Sect. B Beam Interact. Mater. At.* **1985**, *9*, 344–351. [CrossRef]
39. Pomaska, M.; Mock, J.; Köhler, F.; Zastrow, U.; Perani, M.; Astakhov, O.; Cavalcoli, D.; Carius, R.; Finger, F.; Ding, K. Role of oxygen and nitrogen in n-type microcrystalline silicon carbide grown by hot wire chemical vapor deposition. *J. Appl. Phys.* **2016**, *120*, 225105. [CrossRef]
40. Diyatmika, W.; Liang, F.; Lou, B.; Lu, J.; Sun, D.; Lee, J. Superimposed high power impulse and middle frequency magnetron sputtering: Role of pulse duration and average power of middle frequency. *Surf. Coat. Technol.* **2018**, *352*, 680–689. [CrossRef]
41. Anders, A. Deposition Rates of High-Power Impulse Magnetron Sputtering: Physics and Economics *J. Vac. Sci. Technol. A* **2010**, *28*, 783. [CrossRef]
42. Nivedita, L.R.; Kumar, V.V.S.; Asokan, K.; Rajendrakumar, R.T. Growth and Magnetic Properties of RF Sputtered Fe-Ga Thin Films. *Mater. Res.* **2015**, *18*, 946–952. [CrossRef]
43. Jung, C.-K.; Lim, D.-C.; Jee, H.-G.; Park, M.-G.; Ku, S.-J.; Yu, K.-S.; Hong, B.; Lee, S.-B.; Boo, J.-H. Hydrogenated amorphous and crystalline SiC thin films grown by RF-PECVD and thermal MOCVD; comparative study of structural and optical properties. *Surf. Coat. Technol.* **2003**, *171*, 46–50. [CrossRef]
44. Galvão, N.; Vasconcelos, G.; Pessoa, R.; Machado, J.; Guerino, M.; Fraga, M.; Rodrigues, B.; Camus, J.; Djouadi, A.; Maciel, H. A Novel Method of Synthesizing Graphene for Electronic Device Applications. *Materials* **2018**, *11*, 1120. [CrossRef]
45. Li, W.; Yuan, J.; Lin, Y.; Yao, S.; Ren, Z.; Wang, H.; Wang, M.; Bai, J. The controlled formation of hybrid structures of multi-walled carbon nanotubes on SiC plate-like particles and their synergetic effect as a filler in poly(vinylidene fluoride) based composites. *Carbon* **2013**, *51*, 355–364. [CrossRef]
46. Badini, C.; Fino, P.; Ortona, A.; Amelio, C. High temperature oxidation of multilayered SiC processed by tape casting and sintering. *J. Eur. Ceram. Soc.* **2002**, *22*, 2071–2079. [CrossRef]
47. Wilhelm, M.; Wruss, W. Influence of annealing on the mechanical properties of SiC–Si composites with sub-micron SiC microstructures. *J. Eur. Ceram. Soc.* **2000**, *20*, 1205–1213. [CrossRef]
48. Xia, A.; Huizhao, Z.; Li, Y.; Chengshan, X. Formation of carbon nanowires by annealing silicon carbide films deposited by magnetron sputtering. *Appl. Surf. Sci.* **2002**, *193*, 87–91. [CrossRef]

49. Yang, W.S.; Biamino, S.; Padovano, E.; Pavese, M.; Marchisio, S.; D'Amico, G.; Mio, S.C.; Chen, X.; Fino, P.; Badini, C. Thermophysical properties of SiC multilayer prepared by tape casting and pressureless sintering. *Compos. Struct.* **2013**, *96*, 469–475. [CrossRef]
50. Kuenle, M.; Janza, S.; Eibl, O.; Berthold, C.; Presser, V.; Nickel, K.-G. Thermal annealing of SiC thin films with varying stoichiometry. *Mater. Sci. Eng. B* **2009**, *159–160*, 355–360. [CrossRef]
51. Ferrari, A.C.; Robertson, J. Interpretation of Raman spectra of disordered and amorphous carbon. *Phys. Phys. Rev. B* **2000**, *61*, 14095–14107. [CrossRef]

MDPI
St. Alban-Anlage 66
4052 Basel
Switzerland
Tel. +41 61 683 77 34
Fax +41 61 302 89 18
www.mdpi.com

Micromachines Editorial Office
E-mail: micromachines@mdpi.com
www.mdpi.com/journal/micromachines